Strategic Approaches to Managing Technological Change in Sustainable Business Models

Daniele Leone
University of Naples Parthenope, Italy

IGI Global
Scientific Publishing
Publishing Tomorrow's Research Today

Vice President of Editorial	Melissa Wagner
Director of Acquisitions	Mikaela Felty
Director of Book Development	Jocelynn Hessler
Production Manager	Mike Brehm
Cover Design	Jose Rosado

Published in the United States of America by
IGI Global Scientific Publishing
701 East Chocolate Avenue
Hershey, PA, 17033, USA
Tel: 717-533-8845
Fax: 717-533-7115
Website: https://www.igi-global.com E-mail: cust@igi-global.com

Library of Congress Cataloging-in-Publication Data

Names: Leone, Daniele, 1991- author
Title: Strategic approaches to managing technological change in sustainable business models / [edited by] Daniele Leone.
Description: Hershey, PA : IGI Global Scientific Publishing, [2025] | Includes bibliographical references and index. | Summary: "This book aims to speak about the integration of human intelligence with machine capabilities to optimize production and specifically emphasizes mass customization and environmental sustainability"-- Provided by publisher.

Identifiers: LCCN 2025013521 (print) | LCCN 2025013522 (ebook) | ISBN 9798337334660 hardcover | ISBN 9798337334677 paperback | ISBN 9798337334684 ebook
Subjects: LCSH: Management--Environmental aspects | Sustainable development | Industry 5.0
Classification: LCC HD30.255 .L465 2026 (print) | LCC HD30.255 (ebook) | DDC 658.4/083--dc23/eng/20250813
LC record available at https://lccn.loc.gov/2025013521
LC ebook record available at https://lccn.loc.gov/2025013522

British Cataloguing in Publication Data
A Cataloguing in Publication record for this book is available from the British Library.

Table of Contents

Preface

INTRODUCTION

Industry 5.0 represents a significant evolution compared to Industry 4.0, emphasizing sustainability and human-machine interaction. An approach that integrates human intelligence with machine capabilities to optimize production and specifically emphasizes mass customization and environmental sustainability. In this vein, knowledge and innovation are the development engines for companies operating in current environmental contexts based on technological change. Economies of relation and sharing play a fundamental role in the search for competitive advantage.

The transition from Industry 4.0 (focus on digitalization and automation) to 5.0 moves towards collaboration between humans and robots and optimizing resources. Furthermore, Industry 5.0 aims to reduce environmental impact by efficiently using resources and renewable energy. The adoption of advanced technologies such as Artificial Intelligence (AI), the Internet of Things (IoT), and collaborative robotics (cobots) brings not only profit results but, above all, social benefits: we can finally witness an improvement in the quality of work and the environment. Value creation lies in "business knowledge", which must be shared and made accessible to all stakeholders.

THE CHALLENGES

The difficulty of replicating such know-how within organizations has made it appropriate to implement sustainable business models to ensure the continuous exchange of information between company resources. We must hope for policy-makers. For example, in Italy, with the 5.0 transition plan, the aim is to promote the digitalization and sustainability of Italian companies with subsidized investments and tax incentives to enhance energy efficiency and adopt cutting-edge technologies, putting human being at the project center.

In this vein, looking from the past, in his autobiography (My Life and Work, 1922), Henry Ford illustrates his organizational model of production, analyzing the effects of the assembly line on the working class: the assembly line was based on a rigid division of short, repetitive and standardized tasks. His point of view was at the center of numerous social debates that described Fordist thought as "alienating" for the worker. Although this approach has been criticized for the limits imposed on the condition of workers, it has allowed a drastic reduction in production costs, favoring the accessibility of many products that were previously considered inaccessible due to their prohibitive prices. A prime example is the Ford Model T: its price dropped from $950 in 1908 to $290 in 1927, thus allowing its diffusion also among the middle class. Fordism allowed the transformation of the "isolated" machines of the initial phase of mechanization (pre-Fordism) into a real system of machines interconnected with each other and housed within large factories; this process then led to a consequent standardization of production. The increase in company size favored the achievement of more efficient economies of scale, such as to justify investments directed in this direction. After a short time, Big Corporations emerged, large industrial structures capable of sustaining accelerated production rates, the result of a continuously expanding market. The socio-economic change of this period was not only explained by the technological component, but the factor that favored the start of mass production was the birth of a new organizational model, developed by Frederick Winslow Taylor and Henry Ford: a rigid system inspired by the hierarchy of military structures. In this way, entrepreneurial activity was not only structured in a more rational way but also saw the appearance of crucial changes: the search for a more efficient working method led to an internal division between operational work and cognitive work and, therefore, to the separation between manual tasks and planning, coordination and control. A further change concerned the establishment of a hierarchy within the production system, divided into three levels: workers, middle managers and first-level bosses and the management level. Finally, emphasis was also placed on workers' consensus, obtained through economic remuneration and production incentive systems. As anticipated, the Fordist model was the subject of numerous criticisms that identified in the model a total lack of attention to worker protection, which inevitably led to a crisis of the Fordist regime starting in the 1970s, paving the way for a more flexible and decentralized production model. The reasons behind the Fordist crisis lie in the contradictions of the regime itself.

During the years of the Fordist regime, attempts were made to leverage a certain type of capacity: reproducible capacity, i.e. knowledge that is produced to be reproducible in numerous applications at almost zero cost. Consequently, large Fordist companies tried to adopt the logic of maximum vertical integration and dimensional growth, exploiting knowledge as much as possible to increase the volumes of standard products and incorporating small and medium-sized companies present in various sectors. It follows that such a system requires a strong stability of decision-making power at the top, such as to be able to guide the economic and social behavior of workers. In the long run, in fact, the extreme rigidity of the system resulted in complexities that were difficult to control, and therefore, in a crisis of the system itself. The crisis began to manifest itself under multiple uncontrollable dynamics: first, the variation in demand, which became less and less conditionable, the arrival of transnational competition (particularly with Japan) that the large established companies were unable to manage. All this began to result in a monetary instability that radically shook the sales programs, to the point that even the State lost its ability to mediate and direct within the system. It was in this context that the birth of the phenomenon called post-Fordist capitalism (1970-2000) was experienced. The post-Fordist period valorized the new market demand and the search for production and management flexibility, making use of the entrepreneurial initiatives adopted by small industrial, artisan and commercial companies.

In this context, particular emphasis was given to two factors: people and territory, the former allowed an efficient division of labor between subjects with different skills and functions in the supply chains, the territory instead, favored the formation of local innovation systems, and allowed the diffusion of "self-produced" knowledge from experiences in a specific sector of specialization6, favoring an initial approach of entrepreneurs to know-how and subsequently also to the concept of customer satisfaction. This search for flexibility was guaranteed above all by the birth of new digital technologies, which opened the doors to another important phase in the history of innovations, that of digitalization.

SEARCHING FOR A SOLUTION

Solutions to the problem of managing technological change in sustainable business models in the new industrial paradigm hark back at shifting emphasis from technology to business and social process.

During the planning of personal research activity in management studies, technological development has assumed a primary role in studying the factors discriminating against the knowledge transfer of the latter among the members of new organizations. The digitalized era in which we live is characterized and influenced by the largest methods of knowledge distribution, such as digital platforms, social media, networks, and so on. Experiences of mass collaboration, new production mechanisms, and consequent dematerialization of the system output represent constitutive elements of the new "competitive environments" framed in the so-called crowd-based phenomena based on the difficulty of managing technological change. The current socio-economic context in which the new organizations operate must be observed not only as a source of risks but also, above all, as new opportunities. Therefore, it is understandable to think that we are in a moment in which the corporate paradigm is evolving. The firm-centered approach used up to now is being abandoned to move towards a world-centered model, in which most decision-making processes are oriented towards achieving sustainable objectives for the community. Obviously, if something is changing it is because opportunities have emerged. New technologies supporting companies can make a difference in the transition process towards sustainable development.

The concept of sustainability must be understood following 3 fundamental dimensions. First, economic sustainability must be ensured, with the aim of creating an efficient system, capable of generating added value and able to develop new services for people and businesses. Social sustainability must then be considered, with the aim of reducing inequalities by improving various aspects of the community. Finally, environmental sustainability must be ensured to reduce pollution at a global level and to achieve the objectives defined at an international level

Thus, we live in a system in which we have been taught to pay attention to the environment, from school to politics up to companies, involving children, adults, workers, and the elderly. Sustainable development has emerged as a relevant topic in many sectors, involving the interest of politicians, experts, scholars and entrepreneurs from all over the globe who have mobilized to ensure such development occurs.

The case study methodology can be considered a research strategy that aims to study units of analysis, which can be individual subjects, groups, or institutions defined precisely as cases. A case study is a specific approach to research, a project that uses targeted sources, adequate analysis techniques, quantitative and qualitative sampling procedures, and a specific organization of the data collected to preserve the unitary character of the object studied. This study starts from an already acquired basic knowledge and, through data collection and analysis, provides empirical evidence helpful in answering the questions that guide the study itself. Case studies are longitudinal, involving the acquisition of multiple data on the same referents, or transversal, i.e. the acquisition of data on different referents in each period. The data collection techniques used in the case studies are focused on what is necessary and the research design is not rigid but is built based on the specific case to be studied. The sources of evidence can be made up of people, groups, environments, artifacts, and archives; the information collected and processed must be connected in an organic whole without losing relevance, validity, and reliability, showing a clear and engaging overview of the reader as convincing to other researchers.

Therefore, Industry 5.0 aims to create intelligent production plants in which machines and systems can communicate, analyze data and make decisions autonomously, increasing efficiency, productivity and adaptability in production and beyond. To date, there is talk of a strong connection between Industry 5.0 and the circular economy since both aim to optimize the use of resources and reduce waste through technological innovation. Enabling technologies constitute a set of tools and practices capable of optimizing the use and consumption of resources, thus aligning with the principles of Circular Economy. In this context, studies have classified sustainable business models into three categories based on their objectives: prolonging the use of resources by increasing their useful life, recycling as many resources as possible within the value chain and reducing resource consumption for each product. Consequently, the use of enabling technologies of Industry 5.0 is essential to obtain precise data provided with such speed that it is possible to monitor the availability and condition of resources in real-time. The Industrial Era 5.0 has therefore marked the beginning of a technological revolution that is changing the way we live, work and interact. In this scenario, the financial sector has undergone a significant transformation, especially with the affirmation of FinTech (financial technology). FinTech has revolutionized the financial services sector, offering more efficient, inclusive and innovative solutions.

Two important aspects of this evolution are Crowdfunding and Blockchain, which have changed the methods of financing and economic transactions. It has been observed so far, how companies are approaching Industry 4.0 and 5.0 through the adoption of enabling technologies, which contribute to establishing a relationship of trust with the public, ensuring greater transparency and providing investors with more detailed information. Furthermore, having these technologies the ability to "streamline" commercial operations and revolutionize the methods of financing, it has been observed how crowdfunding, today, is considered a more reliable option for the public.

The integration of these technologies has the potential to revolutionize the financial services sector, creating a new paradigm that is more inclusive, transparent and efficient, which is witnessing an increasingly rapid development of FinTech. Crowdfunding, on the other hand, has allowed individuals and small businesses to access sources of financing that were previously difficult to reach, promoting economic growth and financial inclusion. FinTech, therefore, provides the digital infrastructure necessary for the functioning of crowdfunding platforms, improving the security of transactions, transparency, and accessibility, which is why crowdfunding is one of the most relevant applications of financial technologies.

Another business function addressed in this monograph is marketing. The evolution of marketing has led to consumers being considered not as agents who make their choices based on rational information but that, in reality, their choices are the result of influences and are based on the emotional and social part in which they find themselves. Economic sciences are not a science in itself but a set of disciplines that collaborate in order to understand individuals from a quantitative, social, and psychological point of view. In conclusion, it is possible to state that neuromarketing is fundamental to understanding how and where to invest resources to market the products and services produced by companies. The main obstacle that prevents neuromarketing from being used efficiently by companies is the fact that the techniques used are expensive and difficult to apply in restricted business contexts. A further obstacle is found in the fact that the techniques can be invasive, raising issues related to the morality of some techniques, which lead to conditioning the consumer to the point of bypassing him at the time of choice. Despite the criticisms raised, the latter is fundamental and, therefore, should be used where and when possible; it is important to invest resources to make it efficient and accompany the consumer in his choices.

Furthermore, although our findings emphasize the key benefits of Industry 5.0 technologies, this research has limitations. The technologies that enable new digital transformation are, for example, 5G applications in hybrid solutions. Different software agrees to have portable platforms available for radio communications and 4G and 5G are allowing the integration and increasingly innovative introduction of communication networks in a systemic key, with the possibility of significant improvements in terms of bandwidth, coverage and energy and economic efficiency. The goal is to access 5G with the evolution of existing wireless technologies such as LTE, HSPA, GSM and WIFI. The most avant-garde activity in communication networks is that of space. The application field concerns data transfer between satellites and ground stations, from communications between satellites to applications of Internet of Things (IoT) technologies up to Industry 4.0 management. Future research could also integrate quantitative studies making the framework generalizable. Additionally, future research could conduct comparative analysis among different countries or in a longitudinal perspective. Unfortunately, this topic has only become relevant as the years have passed; in fact, initially, this factor was ignored entirely within national economies, and the situation we are currently experiencing is exact proof of this: melting of glaciers, rising seas, holes in the ozone layer, environmental and sea pollution and, consequently, all the diseases that come from it. Safeguarding the environment means safeguarding ourselves and the generations that will succeed us. Managing technological change through sustainable business models means looking to the future without the classic lens of profit orientation but also by looking at corporate social responsibility.

ORGANIZATION OF THE BOOK

In this book I intend to present research on *Strategic Approaches to Managing Technological Change in Sustainable Business Models* divided into eight chapters.

In the first chapter, I will first present some theoretical aspects related to knowledge and innovation management literature to sustain technological change and how knowledge is intended as the basis for competition in the digitalized era. Later, in a second paragraph, I will illustrate which are considered the main theories about the diffusion of innovation so as the reference authors. Finally, I will focus on the types of innovation in technological change management and on the passage from the closed to the open innovation.

In the second chapter, I will deal more specifically with searching for strategic sustainability in digital transformation and business model innovation. In the first paragraph, I'm going to present the importance of digital capabilities in organizations, and in the second paragraph, the most recent studies on the research of strategic sustainability in business models and different perspectives from different authors. Later, I will focus on research aimed at studying the element of business model canvas based on the need to jump to the business model innovation theories. I will deal with the theories of some of the most important and influential contemporary management scholars to analyze what drives organizations to innovate and digitalize their processes and which type of business model they have to apply to achieve a competitive advantage. In the final paragraph I will propose a focus on the role of corporate culture in digital innovation.

In the third chapter, I intend to discuss how organizations manage the value co-creation dynamics in the sharing economy and circular economy. by exploring different case studies related to recent trends in the service markets. We will illustrate different theories and perspectives by discussing a new dominant logic: from product to service following the innovative approach of "servitization" within the actual digitalized and complex contexts.

In the fourth chapter, I will first present some theoretical aspects of neuromarketing strategies to analyze the value co-creation aspects within the business processes of different organizations. This approach has led some researchers and marketing scholars to reflect on the fact that, by analyzing only the decision-making processes related to logical and rational thinking, we would end up neglecting a significant part of consumer behavior. One of the main objectives of neuromarketing is, therefore, to try to fill this gap, offering tools to understand the unconscious mechanisms that influence consumer behavior and choices. For example, emotions, implicit associations, and unconscious automatic processes can be held responsible for most decisions. Finally, in the last paragraph of his chapter, I will illustrate different case studies to in-depth explore the phenomenon.

In the fifth chapter, I will deal more specifically with the difference between the Industry 4.0 and Industry 5.0 paradigms. In the first paragraph, I'm going to present the fourth industrial paradigm and the technologies that act inside, with a specific focus on some "complementary enabling technologies". In the second part, I will show the new Industry 5.0 intended as a cultural revolution where human beings are repositioned at the center of the main strategic firm approach for managing technological change. Finally, different case studies are presented with a deep explorative analysis.

In the sixth chapter, I intend to discuss the innovation of Media franchises: business models and Industry 5.0 Technologies of transmedia content by exploring different case studies related to recent trends in different sectors. In the first paragraph, I'm going to present the understanding of media-franchise from the foundations to the key concepts. In the second part, I will illustrate the role of knowledge in effective media-franchise management. Finally, we focus the attention on technological innovation and consumer-centric marketing on the streaming business.

In the seventh chapter, I will deal with the theories of some of the most important and influential contemporary management scholars for analyzing the evolution of human resource management and which type of digital skills firms have to apply to maintain a positive environment and to create value among workers. In the second part, a specific reference to the employee well-being and an illustration of different theories and perspectives of corporate social responsibility. Finally, in the last paragraph, I will conduct an explorative analysis by showing different case studies of firms involved in various sectors.

In the eighth and last chapter, I will first present some theoretical aspects related to sustainability and technological change management literature and how it is important to manage and acquire knowledge of new technologies to design future paths in our economies without creating disorders and failures in firms' business processes. We will focus our attention on the transformation of existing models to avoid chaos in different service sectors.

Daniele Leone
University of Naples Parthenope, Italy

Section 1
Digital Transformation and Technological Change: How to Manage?

Chapter 1
Knowledge and Innovation Management in Organizations

ABSTRACT

After summarizing the main key studies about the importance of knowledge and innovation management, the chapter reviews the theories, and the models applied to the diffusion of innovation to manage the technological change in the current digitalized era. It illustrates the main characteristics, forms, and application domains. Numerous studies attested that open innovation is crucial to supporting and optimizing business processes in different organizations involved in digitalized ecosystems. Analogies and distinctions between several types of innovations are also discussed. In this vein, knowledge and innovation management can allow companies to understand their potential customers better, going beyond the information provided by traditional market research and understanding the real motivations that drive purchasing decisions, even when these are not immediately evident at a rational level. Therefore, this chapter aims to investigate the inseparable link between knowledge and innovation in today's complex and digitalized environments for managing technological change.

INTRODUCTION: KNOWLEDGE MANAGEMENT AS A BASIS FOR COMPETITION IN A DIGITALISED ERA

The current economic scenario in which companies operate is defined as rather 'turbulent' due to the high environmental complexity, the growing degree of uncertainty, and the speed with which changes occur within the market. This aspect includes, for example, environmental regulations, the management of natural resources, and

DOI: 10.4018/979-8-3373-3466-0.ch001

the impact of production activities on the climate and ecosystems. In order to remain competitive, companies must not only comply with increasingly stringent laws but also adopt sustainable practices, such as the use of natural ingredients and recyclable packaging. In this way, in addition to reducing their environmental impact, they also respond to the growing sensitivity of consumers towards eco-friendly products and sustainable business models (Geissdoerfer et al., 2018), improving their image and contributing to the protection of the planet.

In the era of globalization, every company must deal with increasingly vast markets, with numerous differences from an organizational and strategic point of view. This entails a great deal of difficulty in identifying effective strategies that can make the company truly competitive in a market in which, by now, competitiveness has grown exponentially. For this reason, companies will have to make innovations a real strategic weapon.

Recent studies have shown that knowledge is the most effective tool that companies have at their disposal to obtain a competitive advantage that can be preserved in the long term; among these, we recall the studies conducted by Nonaka and Takeuchi (1996), who first modeled the process of knowledge creation by placing it at the center of the governance and organizational structure of the company, or more recently the contribution of Teece (2000), who argued that the most strategic sources of competitive advantage are related to the development and use of knowledge assets.

Traditional strategies were based on a 'primitive' concept of knowledge resulting mainly from the combination of resources/skills within the company and threats/opportunities arising from the environment outside the company. In the Fordist era, for example, information was difficult to transfer as knowledge was exchanged only between the members of the 'knowledge core' within the company.

Today, the information and resources owned by the company are the object of study and processing by the members who will have to use them for their best use.

It is necessary to make a distinction between knowledge and other productive factors, and it should be specified that simple possession of knowledge does not ensure the company obtains an advantage; it's rather necessary that the use and use of the same is intentional by the enterprise itself.

In the past, companies did not have a strong need to optimize knowledge management by they operated in a stable environment that allowed the normal use of traditional production factors while still achieving good performance within the market. Knowledge enhancement has increased until it marks a fundamental step in economic managerial theories. We can, therefore, apply to this scenario a statement by Nonaka, who, since the 1990s, contributed to the literature on the subject, arguing that: "in an era in which the only certainty is uncertainty, the only sure source for the competitive advantage is knowledge. "

Before understanding what elements, knowledge is made of and what the different types exist, let's define what knowledge means. From the different interpretations provided by scholars, we can define "knowledge" as the awareness acquired over time resulting from a continuous learning process based on efficient data and information combined with individuals' skills, ideas, and skills.

We can, therefore, affirm that knowledge is closely linked to the individual's experience, contributions, and abilities, such as the ability to evaluate, judge, values, and culture. In addition, it is formed by a dynamic process that begins with acquiring data and then transforms these into information and knowledge.

In the specific business environment, knowledge is represented by intellectual capital, which constitutes the company's most valuable resource for creating and sustaining a competitive advantage. Knowledge within the company took on primary importance only in the late 1980s when it was identified as a "critical success factor for the pursuit of competitive advantage" (Itami and Rohel, 1991).

In the Fordist period, there was a market where "knowledge" was transferred only through direct contact between members of the same company and within the so-called "knowledge core"; therefore, information was difficult to transfer (Golinelli and Dezi, 1997; Dezi et al, 2019). Nowadays, there is an increasing flow of knowledge from country to country.

These premises help to undertake an analysis of knowledge as a competitive resource for the company; before this, it is necessary to deepen and clarify the concepts and relationships that exist between resources, general skills, and business skills to understand and frame the concept of knowledge as a dynamic and non-static resource within digitalized contexts. For instance, the industry 5.0 context makes available technologies like AI in fashion which improve the efficiency of different operations.

AI makes a huge contribution to sustainable product design, integrating eco-friendly materials and low-impact production processes at all stages of the product life cycle (Leone et al., 2021). AI-driven garment creation reduces the environmental impact of the fashion industry: the combination of technology and sustainability has revolutionized the industrial field, giving rise to Industry 5.0.

The fifth industrial revolution focuses its pillars on the human being, sustainability and collaboration, laying the foundations for a more resilient, inclusive and environmentally friendly industrial future.

Unlike previous industrial revolutions that focused their interest on automation and maximizing production, Industry 5.0 favors the circular economy, reducing waste and the environmental impact of production processes. The circular economy is an economic and production model that aims to minimize waste and maximize resources. The life cycle of products is thus extended, helping to minimize waste: once a product has finished its function, the materials it is made of are reintroduced, where possible, through recycling and generate further value. The circular economy approach contrasts with the goods offered by Fast Fashion, characterized by short life cycles notoriously associated with the concept of a linear "use and throw away" economy. This latter model divides the life of each product into five phases: extraction of virgin raw materials, production of consumer goods through the use of energy and work, distribution of products to consumers, consumption of the same, disposal of the "scraps" that have become waste.

Scholars have addressed the theoretical-conceptual debate on analyzing internal company resources since the 1990s according to the *Resource-based View (RBV).*

The detachment from the studies of the Industrial Organization has determined new levers for companies to construct a lasting and sustainable competitive advantage. The attention is thus directed to the activities and processes of evolution of internal resources to which the skills and competencies on which the resource-based approach is based are connected.

Therefore, a causal relationship exists between the heritage of resources, the skills, and the company's competitiveness. The objective of entrepreneurial conduct of the new realities is understanding a transition from resources and skills to the skills that are determined in integrating and reconfiguring one's portfolio to sustain the competitive advantage.

Furthermore, in a complex and constantly evolving environmental system, companies are more exposed to the entry of potential competitors. The aim is to seize opportunities and defend against threats. Still, all this presupposes a high capacity of managers to create the necessary and sustainable conditions for the interaction of the company with its environment.

What has just been said is in harmony with the vision of Penrose's business (1959), which is a set of resources that must be exploited optimally. Selznick (1957) is one of the first to theorize the concept of "distinctive skills" as determinants of the success of organizations' growth strategies.

Therefore, the company must be understood as a heterogeneous set of resources and skills, which are not limited to mere availability but are the subject of study by the actors who must be able to combine them in the best way. Furthermore, suppose the company possesses skills with a "unique" character. In that case, it means that the so-called advantages that have been achieved are firm-specific, which creates the basis for the pursuit of a lasting and defensible competitive advantage. The RBV confirms all this.

In such a context, the company is considered a system characterized by parts and sub-systems (people, resources, processes, units) that must be harmonized in the relationships that occur. Once the business perspective as a system of parts and sub-systems has been clarified, it is appropriate to analyze the business theme as a system of resources, skills, and capacity.

Based on the degree of accessibility, we can distinguish knowledge into:

- *Tacit knowledge* (in the human mind): it is difficult to access because it is strictly personal and subjective. In this case, the difficulty lies in transmitting it and communicating it to the other company members.
- *Explicit knowledge* (in documents): it is easily accessible because it is encoded and processed in documents. The problem that could arise is managing an overload of knowledge that could be an obstacle to business activity rather than a success factor.

Concerning this distinction, we could quote Michael Polanyi's contribution in the text 'The Tacit Dimension' (1966) starting from the statement 'We know more than we know how to say' in the sense that the explicit dimension of knowledge is accompanied by it is always based on a previously internalized tacit dimension. From this, it is possible to understand the importance of tacit knowledge, so much so that to enhance it in the business environment, some scholars propose creating a social system through which members can confront each other and dialogue to expand and share their knowledge and goals.

On the other hand, if one were to consider only explicit knowledge, a big mistake would be made because, as demonstrated, this represents only the tip of the cognitive iceberg of the company structure.

A further distinction of the types of knowledge can be made based on the strategic support it can provide to the company, that is, based on the value it brings to the organization (Cavalli, 2008).

- *Basic knowledge* represents the minimum level of knowledge necessary to carry out the activity; it is present in all companies operating in the various sectors/markets.
- *Advanced knowledge* allows the company to maintain a competitive position concerning companies operating in the same sector with different knowledge content. In this case, the company has a double strategic choice: compete with other companies hoping to have more knowledge or differentiate their knowledge.
- *Innovative knowledge* is possessed by companies that occupy a leading position in the sector; in this case, the power of the leading company can also consist in changing market rules.

Thanks to this distinction, each company can better understand the knowledge 'gap' compared to other companies in the same sector to reduce it and aspire to a better competitive position.

Nonaka and Takeuchi, Japanese economists and professors, have contributed to theorizing the transition from knowledge as pure research to knowledge that spreads and influences managerial processes; in fact, they argue that the real success of Japanese business organizations is based mainly on the ability and experience in creating different cognitive forms.

The 1995 elaboration by Nonaka and Takeuchi, subsequently published in the text 'The Knowledge-Creating Company', is important among economic-managerial theories. These aim to formulate a theory on the creation of knowledge.

First, the two authors specify the distinction between information and knowledge: the latter is closely linked to the individual's values, beliefs, and points of view. Secondly, they affirm that knowledge, unlike information, is directed to an end and is connected to a practical and real situation. Finally, both depend on the interaction that exists between the individuals in the organization and relate to the reference context.

Nonaka and Takeuchi have developed a model of organizational knowledge creation based on two dimensions: the epistemological dimension and the ontological dimension. From an epistemological point of view, Nonaka takes up Polanyi's theory and distinguishes explicit knowledge, i.e., the tip of the iceberg, easily identifiable and transferable, from the tacit one, i.e., the submerged part, which, however, represents the true source of innovation and has greater strategic value within the company.

On the ontological level, however, Nonaka states: “Knowledge is created only by individuals, and an organization cannot create knowledge without individuals; rather, it must support their creativity and provide a context in which they can create knowledge. " From this perspective, we can deduce that knowledge requires a diffusion process from the individual to the organization.

The process of 'knowledge conversion' is the one that generates organizational knowledge that from the individual expands to the organization through the interactions between tacit knowledge and explicit knowledge. Knowledge can, therefore, circulate in the organization according to this “conversion,” which is essentially a process based on the interaction between individuals.

From the 'spiral of knowledge' model developed by Nonaka and Takeuchi, it is possible to identify four types or stages of conversion based on the knowledge that is transmitted and that which is created:

- Socialization
- Outsourcing
- Combination
- Internalization

Socialization is the process of converting from tacit knowledge to tacit knowledge. This type of knowledge is transferred through direct contact between individuals; for example, learning a manual job can also take place without the use of language, and learning can occur through observation, imitation, and practice.

Socialization occurs more frequently in the business environment when collaborations or joint ventures between companies are established; in this case, different knowledge is assimilated tacitly thanks to the work operators of other companies.

Outsourcing is the process of converting from tacit knowledge to explicit knowledge. This is the moment in which knowledge is made explicit, and the codification process takes place, but it is also the most complex moment since tacit knowledge cannot be codified, it is intrinsic to the individual and difficult to formalize. Metaphors and analogies are often used to facilitate this conversion process. Using these forms allows individuals, through figurative language and imagination, to bring out what is acquired tacit knowledge.

Combination is the process of converting from explicit knowledge to explicit knowledge. In this case, the conversion is easier as the knowledge is already encoded and is easily transferable. Individuals can exchange knowledge through communication media, documents, and networks; the possibility of combining and reconfiguring explicit knowledge can generate new forms of knowledge.

Internalization is the process of converting from explicit knowledge to tacit knowledge. This process consists in internalizing and making the codified knowledge one's own, thus passing from learning to action (learning by doing). Internalization enables awareness of different know-how to be learned through the other conversion stages.

The four phases described must never be considered separate; they must necessarily be interpreted with a view to mutual complementarity and interdependence; precisely for this reason, Nonaka and Takeuchi define them in a spiral.

The two authors identify favourable conditions that create a suitable context for the spiral to start and evolve. There are five key conditions:

- Intentionality
- Autonomy
- Creative chaos
- Redundancy
- Minimum variety required

Intentionality, or organizational intentionality refers to the organization's aspiration to achieve predetermined objectives. One of the values that must guide workers is a common vision that can make sharing more favorable.

Autonomy is considered freedom of expression by individuals in their work. This, however, must be kept distinct from isolation, as in any case, participation and collaboration between individuals is necessary to produce new knowledge.

Creative chaos originates from a fluctuation, that is, from a break from routine and organizational habits in which situations are characterized by poor predictability. A fluctuating situation can occur in times of crisis in which ambiguous signals also arrive from the environment, in this case, the organization is forced to interact more with the external environment in search of new solutions and new points of reference. Creative chaos is, therefore, a situation that can arise independently of the relationship between the organization and the environment or can be induced by management.

Redundancy understood as the redundancy of information possessed by individuals, allows a multiplicity of people to deal with the same problem with a greater probability of finding effective solutions. This way, we promote the sharing of tacit knowledge and create new communication channels.

The minimum variety required refers to the necessary harmony between the organization and the external environment; the members can respond to changes only if they possess sufficient variety. Variety can be created by combining information in different ways with speed and flexibility, in fact the greatest variety will occur when each member can access a wide range of information through the fewest possible steps.

The results of the resource-based studies tend to consider an economic context where companies with heterogeneous resources also conform differently within the same sector groups and hold these differences over time thanks to imperfect mobility and non-replicability of resources.

In addition, the reasons for business success come from internal elements with certain attributes concerning competitiveness and knowledge sharing (Leone and Schiavone, 2019). The motivations refer to the growth needed to achieve sustainable and defensible positions over time.

In the first paragraph, the main features, classifications, and developments of resource-based studies on resources, skills, and abilities have already been addressed. It has been possible to denote that while the former's conceptual framework presents convergent opinions regarding the capabilities and competencies, there are still conflicting and ununiform opinions.

The various stages of development of skills and capabilities among the company members are glued together by a vital intangible resource, which is "knowledge".

Nowadays, intangible resources create diversity in the performance of companies and, therefore, fall under the utmost consideration to achieve competitive advantage.

In the late 1980s, Itami (1991) identified knowledge as a new basis for competition, and from that point on studies will begin identifying said concept as the new critical factor for success.

Knowledge-based organizational realities define the necessary actions for integrating and applying knowledge as the primary bases for obtaining profits. Another objective is to ensure the correspondence between the company's identification as a production unit and an exchange structure. The active members within the organizations develop different knowledge, the company must be observed as a place of integration and coordination of the specialist knowledge of the individual actors.

Knowledge-based companies aim to increase knowledge and integrate it into goods and services to be brought to the market.

In summary, it can be said that knowledge is intended as a business resource according to the Resource-based theory and used by individual members of the organizational realities to produce goods and services according to the Knowledge-based perspective (KBP).

The tool of knowledge begins to assume a primary role concerning the other productive factors; it is not easily quantifiable; it can self-generate and self-renew depending on the various members and application contexts.

In the knowledge-based line, the main economic-managerial changes lie in improving knowledge transfer and processing and conserving data and information.

The space-time dimensions of individuals are reduced, and technological innovation begins to become a constant without interruption.

The Knowledge-based perspective has its roots in the search for competitive advantage by means of integration actions between internal and external resources.

The content of the business activity concerns the creation and management of the so-called knowledge assets; therefore, we begin to feel the need to better manage (Knowledge Management) all the components that concern business know-how. Companies also observe the external environment as a source of opportunities and not only of risks, but the goal is also to exploit and be proactive to sudden changes in the reference socio-economic context.

Often, we hear people discuss how a certain form of knowledge can be perceived and from this, it can be said that the business objective is to provide a holistic view of the know-how, therefore, to make the most of the KBP, one must be able to consider the types and processes through which knowledge is formed, transferred and used.

The different roles of knowledge in the business environment have made studying it in its many perspectives and facets possible. Managerial studies have addressed the issue of knowledge concerning organizational learning, the behavior of the actors, knowledge management, and the concept of knowledge as a strategic resource.

The goal is to focus on *Knowledge Management* and its impact on companies operating in the current socio-economic context.

The first approach to this discipline dates back to 1986, the year in which Karl M. Wiig coined the term; over the years many economist writers have reflected in depth on the importance of knowledge management in the company, up to 1995, the year in which Nonaka and Takeuchi publish the book The knowledge-creating company: how Japanese companies create the dynamics of innovation. From this moment on, the vision of Knowledge Management changes radically: it becomes an indispensable element for many companies.

It is precisely through Knowledge Management that we try to manage the 'knowledge' resource as a real competitive resource. At this point, we can define Knowledge Management as a managerial response to the growing importance of knowledge. We can identify it as that process that uses and consciously manages knowledge within an organization in order to achieve competitive positions in the context in which it operates.

The primary objective of knowledge management is to provide knowledge workers (those who make daily decisions within the strategic business structure and who will decide their subsequent success or failure) with all the intellectual capacity of the company. In particular, there are three indispensable variables that knowledge management practices must take into account:

- *People*: they are holders of knowledge, especially the tacit knowledge that needs to be made explicit in the organization, and it is on this that management's efforts must be concentrated more to ensure proper knowledge management.
- *The Organization*: knowledge management on the one hand strives to make knowledge accessible, but on the other hand, the value of knowledge is incomprehensible outside the context in which it developed. Therefore it is necessary to understand which information dissemination processes can be adopted and which business processes can allow a fair compromise between these two contradictory trends.
- *Technology*: it can be considered a tool to support knowledge management, as it favors the transmission and sharing of information.

Therefore, knowledge management should be understood as a system of methodologies and processes to manage better the identification, acquisition, dissemination, conservation, and enhancement of knowledge to improve business activities.

Thus, *Knowledge management* is a process of identifying, acquiring, and exploiting collective knowledge within an organization to achieve competitive positions in the contexts in which it operates.

Lundvall and Jhonson (1994) frame the concept by paying attention to the critical aspects of organizational learning, survival, and the continuous evolution of the external environment.

Knowledge management can be understood as a process that aims to identify a synergy between the information learned and processed and the innovation capacity of individuals.

The assumptions of Knowledge Management, therefore, lie in various activities:

- the goal of value creation;
- strategic planning for the management of social capital and intellectual capital consistent with the achievement of the competitive advantage;
- identification of distinctive skills, sharing of information, and control of the wealth of knowledge acquired by members of the organization;
- initiation of evaluation and decision-making processes in order to organize the creation of knowledge over time.

After defining and analyzing the main components of Knowledge Management, we identify some phases of knowledge management theorized in 'Knowledge and Management'. How to enhance the company's knowledge assets " (Cavalli, 2008):

- The Discovery concerning the identification of the sources of knowledge.
- Distribution, ie the link between sources and authors.
- The Acquisition refers to the connection and sharing of knowledge.
- Creation identifies the development of new knowledge for the benefit of the community.

In this process, it is necessary to specify the importance of structuring one's knowledge, defined as knowledge assets; the extraction and structuring allow the company to possess multiple pieces of information thanks to the enhancement of the available data.

This occurs with processes that allow the identification of relevant data, the use of data recovery tools, the formulation of data in a communicable form, archiving, system implementation and finally the development of communication systems and sharing of knowledge.

The other fundamental step concerns the need to develop and use the knowledge assets in the various company areas, to carry out this process are necessary: the recognition of the importance of knowledge on the part of management and stakeholders, the creation of a business environment oriented towards sharing, collaboration and mutual trust, the development of a strategy that indicates the type of knowledge to be used, the search for common values and the creation of a common corporate language, the understanding of the process that will guide the sharing of knowledge and finally the meeting between those who possess knowledge and those who need to know.

As mentioned, knowledge management impacts three fundamental business variables, so interventions aimed at promoting knowledge enhancement must necessarily integrate with people, organization and technology.

The interventions on the organizational structure may concern the entire organizational structure, business processes, the roles of the subjects in charge of knowledge management and the interaction processes for the exchange of information and knowledge between individuals.

The interventions on human resources are carried out to create a corporate culture based on sharing and collaboration, therefore, these can be implemented to implement or modify individual and collective skills, models of interaction between people, evaluation and incentive systems for the exchange of knowledge, the in-depth analysis and updating systems.

As regards technology, an indispensable element for sharing information, it is advisable to have some elements such as networks, search engines and information systems that can simplify the exchange, collection and distribution of knowledge.

From the interventions carried out on the three variables of knowledge management, we can identify three categories, including the phases of the knowledge life cycle.

- The Knowledge Repository includes knowledge storage and organization activities.
- Knowledge Sharing concerns transfer and sharing through the appropriate communication systems.
- Knowledge Discovery is about developing and using new knowledge.

Figure 1. Main categories of knowledge management (Author research)

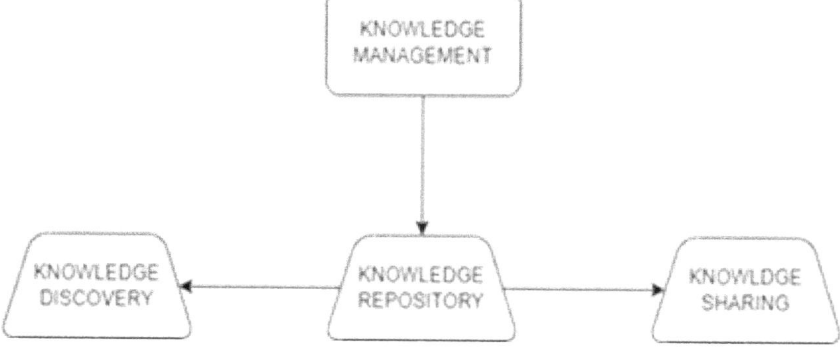

Each knowledge management plan is based on dialogue and comparing the various organizational units. To simplify this management process, it is necessary to introduce figures to whom professional roles, functions and responsibilities are assigned.

We can identify four main functions that are indispensable for knowledge management:

1. *Knowledge Oriented Personnel*: it refers to the knowledge of all workers, workers or employees, who, in carrying out their work create, distribute and use knowledge.
2. *Knowledge Management Specialists*: This category includes knowledge management professionals such as knowledge engineers who acquire, infer and structure knowledge. In fact, investing experienced and qualified people in this area favors a correct enhancement and organization of knowledge.
3. *Knowledge Project Managers*: subjects entrusted with the responsibility of the knowledge project. The main tasks of this function concern: project development, management of customer expectations, and identification and resolution of project problems.
4. *Chief Knowledge Officer*: is the figure of the top manager of knowledge. The person in charge of this function must be able to design and supervise the supporting elements of knowledge (research centers, networks, libraries), maintain relationships with external suppliers of information and knowledge, collect, measure and define the value of knowledge and must have the skills to create a community among individuals and direct them towards collaboration. IT (Innovative Technology) skills and deep and detailed knowledge of Knowledge Management practices are required to fill this function. Usually, the Chief Knowledge Officer is a high-level manager coming from an internal experience of the company and therefore a deep knowledge of its dynamics and the cultural climate present in it.

The events that have stimulated greater reflection and greater insight into the topic of knowledge management concern: are the forces exogenous and endogenous to the company. Among the exogenous ones we can identify: the globalization of markets which has widened the competitive basin; the evolution of the needs of consumers who are increasingly informed about the goods they buy and are therefore always looking for goods that can best meet their expectations; the evolution of suppliers and the increase in technologies that the company has available. Among the endogenous ones are the growing awareness of the importance of cognitive resources, which generate a competitive advantage through learning and knowledge management; the displacement of competition on the upstream phases that require a greater intensity of knowledge than the downstream ones.

Therefore, careful knowledge management can help the company react to these forces from the internal organization's point of view and the company as a whole.

From an internal perspective, the advantages deriving from Knowledge Management processes can be summarized in two fundamental points:

- *Better management of business activities.* Companies that adopt a system based on Knowledge Management are able to improve the quality of processes and work activities, thanks to the growth of individual and group skills and thanks to the transmission of these. Furthermore, this circulation of information and knowledge allows for greater speed in solving problems, implementing programs, and designing new products/services on the market.
- *Better management of human resources.* As previously mentioned, knowledge management practices provide for the dissemination of knowledge, so that, in such an organizational context, the sharing of professional experiences facilitates the enhancement of critical knowledge, indispensable for the improvement of the company and for the achievement of goals. Furthermore, the collaborative climate stimulates the involvement and motivation of the staff.

However, from the point of view of the company as a whole, the advantages associated with effective knowledge management concern above all:

- *Greater vision of new business opportunities.* Companies that adopt Knowledge Management methodologies can improve their understanding and study skills of the markets through a systematic survey and to understand trends by reading historical memory. All this makes it possible to predict and anticipate the market's evolution, identify its customers, and propose innovations in the products/services offered in the sector to which they belong.
- *Better responses to market needs.* Knowledge management allows the exchange of knowledge between the various company functions, facilitating the frequency of launching new products/services; thanks to the exchange of experiences and best practices, the sales networks are managed more effectively. Furthermore, the ability to interpret market developments and consumer needs allows companies not only to improve the degree of success of products/services placed on the market but to increase customer satisfaction.

As seen, through a process of extraction-structuring-development-use, knowledge management affects the organizational structure, processes, and above all the methods of management and development of human resources; this approach allows to obtain significant advantages both on the internal front and on the outside the company.

In conclusion, we can say that with better management of the company and its human resources, eater clarity of the opportunities offered by the competitive context, and a better response to the needs of the market, knowledge management is able to increase the overall value of the agency.

The adoption of digital technologies is triggering a real revolution in business processes, paving the way for a new era of efficiency and innovation. This profound change is manifesting itself in various operational areas, radically transforming the way in which companies organize themselves and interact with the market.

In the field of human resources management, digitalization has revolutionized recruitment thanks to sophisticated systems based on Artificial Intelligence. These tools analyze and select candidates quickly and precisely, while collaborative platforms improve internal communication, promoting a more dynamic and synergic work environment.

At the same time, marketing strategies have taken on a new face. The advent of social media and digital advertising platforms allows companies to monitor consumer behavior in real time, personalizing offers and messages in an extremely targeted manner. This data-driven approach not only improves the effectiveness of campaigns but also allows for closer and longer-lasting relationships with the customer.

In the e-commerce sector, digitalization has given rise to highly integrated ecosystems, in which each phase of the sales cycle (from warehouse management to final delivery) is optimized thanks to the use of advanced technologies. This integration allows quick responses to market needs, ensuring transparent, efficient and constantly evolving processes.

Digital transformation is not only a strategic choice, but a necessity for companies that want to remain competitive in an increasingly connected and dynamic world. Digital innovation is redefining the boundaries of what is possible, making companies more agile, efficient and ready to seize the opportunities of the future.

DIFFUSION OF INNOVATION: THEORIES AND MODELS

Traditionally, diffusion is understood as the process by which an innovation spreads in certain communication channels between a certain social system member over time (Rogers, 1962). The term innovation refers to any idea perceived as new by the members of a system.

The theory of the diffusion of innovation is part of the branch of communication theories and management theories to understand how to manage technological change through sustainable business models. It has been extensively studied in the literature from the point of view of different disciplines and regarding different goods, services and ideas. One of the first scholars who paid particular attention to the diffusion of innovation was the French sociologist G. Tarde (1903), who found that the diffusion of ideas takes place according to an S curve, which allows one to distinguish three phases (innovation, growth, and maturity). If time is placed on the abscissa axis and the number of adopters on the ordinate axis, it is clear that in the first phase, innovation encounters many obstacles to spread; in the second phase, namely that of growth, it tends to spread rapidly, until it reaches the third, that of maturity, where the market becomes saturated and the pace of diffusion slows down. Subsequently, B. Ryan and N. Gross (1943), carried out empirical studies focusing attention on the adoption of hybrid seeds, new machinery, and new techniques by independent farmers. The research then opened up to other sectors, including communication, marketing, and the medical field. The two theorists, therefore, configured the diffusion of innovation as a social process, in which the subjective evaluations of entrepreneurs are at the center.

They identify five categories of subjects based on their attitude toward innovation:

- innovators (innovators);
- early adopters;
- the early majority;
- the late majority;
- laggards.

The model of diffusion of innovation was promoted and made famous, however, by the American sociologist Everett Rogers in his "Diffusion of Innovations", (1962).

Before explaining the Rogers model, however, it is right to provide a useful historical reference framework to understand how we have come to speak of the diffusion of innovation today.

The theories of economic development of "neoclassical" origin (or, as it is often said, mainstream) considered technological innovation as a simple factor in a production function.

Karl Marx in his “Zur Kritik der politischen Ökonomie” (1859), recognized the key to economic development in technological progress. According to the German economist, in fact, the competition between the actors pushes the latter towards forms of development such as technological progress.

The role of innovation in the economic world was studied in depth by Schumpeter (1934), who focused in particular on the figure of the entrepreneur.

The Schumpeterian entrepreneur can be understood as the true architect of innovation since he is able to introduce the invention into the production process.

Schumpeter's model formalizes three steps of the innovative process:

1. *invention*, which concerns the results of research activities;
2. *innovation*, which occurs when the invention is applied to the production process.
3. *imitation*, which concerns the attitude of other entrepreneurs who imitate the first innovator forcing him to apply himself again in the search for innovations.

In the Schumpeterian tradition, innovation is nothing more than the reorganization of production factors into new combinations, thus making an axiom the idea that the entrepreneur and the firm must always act with a view to innovating by trying to undertake strategies that can make obsolete the pre-existing production process.

The scholar defines innovation, stating that it has the characteristics of a “creative destruction” force (Schumpeter, 1934).

It determines the future of new activities and productive sectors, with the consequent crisis of the old sectors and technologies. Furthermore, the nature of the same innovative process, as described by the Austrian economist himself, based on the invention-innovation-imitation model, is at the basis of economic growth and above all the instability of the system. In fact, innovation is introduced by a few initial innovators and is widespread in the system by imitators. Schumpeter describes this phenomenon as the 'bandwagon effect, also known as the 'bandwagon effect. He wants to understand that the instinct of the mass is the fulcrum of the dynamics of society. If there is a feeling of approval about a new idea, the crowd will almost certainly follow it. It is a subtle form of manipulation that, however, can modify and reform economic, political, and social systems. A flash example could be that of political polls.

Now that we have a fairly clear historical picture, it is possible to define in detail what he means by diffusion of innovation.

As mentioned in the first lines of this paragraph, Everett Rogers is undoubtedly the main exponent of the theory on the diffusion of innovation. At first, he took up the model that Ryan and Gross devised to explain the distinctive characteristics of each of the five categories.

He showed, for example, how early adopters were more integrated into local communication mechanisms and had a greater ability to assume the role of opinion leaders.

For this reason, Rogers decided to identify the diffusion process as essentially of a communicative nature, in which characteristics and orientations of a subjective nature come into play.

Rather than an S-curve (Tarde, 1903), Rogers appears to be a "bell". For Rogers, the group of innovators is characterized by a high level of education and a propensity for risk. In addition, they are exposed to multiple sources of information. Early adopters, such as innovators, hold high levels of education and enjoy a high community reputation.

The early majority is characterized by individuals who form so-called "peer groups", who often hold leadership positions and go through a deliberative process before considering a new idea.

On the other hand, the late majority is made up of subjects who are normally skeptical, traditionalists, with low economic status, almost totally unwilling to take risks and who are greatly influenced by the social "pressure" exerted by peers; Finally, laggards are composed of individuals who are normally isolated, suspicious, with reduced social relations (only neighbors or relatives), with a slow decision-making process and with limited resources.

The clear distinction made by Rogers regarding the successful factors of innovation is particularly important. Said distinction can be identified:

- *Relative advantage*: innovation must be perceived as better than the others. The degree of "relative advantage" can be measured in economic terms.
- *Compatibility*: innovation must be possible in relation to company standards.
- *Complexity*: concerns the degree to which an innovation is perceived as difficult to understand and communicate.
- *Observability*: innovation that produces visible results has a greater possibility of diffusion.

Rogers' thesis has aroused various criticisms in many scholars. For example, G. Moore (1991) applied the bell curve to the high-tech technology market. According to Moore, Rogers understands the technology adoption life cycle as if it was a continuous process, without jolts or interruptions.

On the contrary, Moore, notes a hiatus (chasm), an interruption, between the early adopters (redefined visionaries, who have high expectations and are inclined to innovation) and between the early majority (redefined pragmatists, who are little inclined to risk and do not have high expectations).

Many innovations in the field of technologies "run aground" precisely in the passage in which diffusion must begin to involve pragmatists (MacVaugh and Schiavone, 2009).

TYPES OF INNOVATION IN TECHNOLOGICAL CHANGE MANAGEMENT

In a constantly evolving market like the current one, it is not enough to successfully manage your company: you must combine strategic and technological innovation to manage technological changes. Industry 5.0 makes available innovative technologies like AI, and companies must react by developing innovative products and processes based on this change. This requires an in-depth consciousness and digital skills of the different actors who have to use these technologies in different business processes.

There are multiple definitions of strategic innovation that all seem to converge on the fact that it materializes in the definition of a new business model. through the development of:

- *Innovative products* (goods/services) presented or combined in a new way to create a radically different experience for customers, involving them also on an emotional level.
- *Innovative processes* for producing and distributing existing or new products that can lead to acquiring new customer groups.
- *Innovative value chains*, to create a new market space that, making the competition irrelevant, allows for an increase in value for both the company and the customer.

The first step to start a strategic innovation is to answer the following questions for the redefinition of the business model:

- Who are our customers?
- What should we offer them?
- How can we offer it to them efficiently and effectively?

Strategic innovation is more effective when combined with technological innovation, understood as a type of process that leads to the expansion and improvement of techniques and product production and distribution technologies. This can also concern concepts other than technology: a better production and distribution system organization is also considered an innovation.

According to the classification proposed by the economist Joseph Schumpeter, technological innovation is distinguished in:

- *Product innovation*. It refers to the introduction on the market of a new product or a new service. It can be a radical innovation if there is the introduction of a new product/service on the market or incremental if it concerns the improvement of an existing prproduct/service.
- *Process innovation*. It concerns introducing a new production process capable of increasing production efficiency, reducing production costs and increasing productivity. Again, innovation is radical if it concerns the introduction of a new production process or incremental if a pre-existing process is improved.

The most successful companies can exploit innovation by acting both on business, organizational, and management models, and on technological advancement, trying to combine these elements to create innovation: this is the logic of modern economies.

The management of technological innovations in companies is a topic that attracts the attention of many scholars. It is stated, in fact, that innovation is the result of an invention or an idea. It, therefore, presupposes that it hides behind it a process that can guarantee positive and sometimes beneficial results, compared to obsolete technology. However, innovation does not always bring the expected results, leading to what is called "technological regress".

One of the first economists who spoke of innovation is undoubtedly Joseph A. Schumpeter, who in "Theory of Economic Development" defines development as a distinct phenomenon, extraneous to what can be observed in the circular flow and in the trend towards balance. Innovation is crucial for industrial change as it destroys a competitive economic context to create a new one. This phenomenon is called "creative destruction" (Schumpeter, 1934).

Affirming as a postulate that there is a unique form or type of innovation is a fairly serious mistake. Scholars, economists, and sociologists have supported the thesis that innovation is such a complex phenomenon, so much so that it is possible to grasp different categories within it.

Abernathy and Utterback (1978), have devised the following model that allows you to follow the life cycle of a given technology at any time.

Based on this model, we can distinguish between two different life cycles: the product technology life cycle and the process technology life cycle.

In both cycles, it is possible to deduce three phases. About product innovation, in its initial phase, a company aims to maximize its performance, which flows into the second phase in maximizing sales (center of the x-axis) up the last phase, which, on the contrary, tends to minimize costs in a so-called "educated" market.

As for process innovation, its cycle is at most asymmetrical. In fact, we start from an uncoordinated production process, to arrive in the second phase of market segmentation. Finally, when the market has stabilized, we arrive at a phase of extreme specialization of the individual parts of the process.

In short, at first, you have a new product that enters the market and that only finds some "early adopters" bringing with it a little standardized production process; then, once the product has become known and the demand has become higher, there is a process standardization.

There is no doubt that not only do these types of innovation exist, but several theorists have also given particular attention to the experimentation of a taxonomy of innovation. One of these is Paul Trott (2012), who identifies further types, which departing almost diagonally from the pre-existing entrepreneurial ideologies and contain a new model of doing business in their most intrinsic side.

- *Product innovation* refers to the development of new products or the improvement of those already on the market, through the use of new materials and technologies or the innovative application of existing materials and products to introduce new ones. performance and meet specific user needs.
- *Process innovations* are changes to the procedural techniques in which a company carries out its activities. Often, they aim at improving the effectiveness and efficiency of the company's technologies. In this case, development concerns a new production process. This innovation can be radical (creating a completely new process) and incremental (improving a process already existing in the company).
- *Organizational innovation.* This may involve a new organizational division, a new internal communication system or the introduction of a new accounting procedure.
- *Management innovation.* Over time, various management innovation systems have distinguished themselves, such as TQM (Total Quality Management) or BPR (Business process re-engineering).
- *Production innovation.* Phenomena such as JIT (just in time) are part of this innovation.
- *Commercial/marketing innovation.* It concerns new methods of financing or new approaches to sales, also developing new company figures such as the marketing director.
- *Service innovation.* It concerns new types of financial services.

Paul Trott is not the only one who divides innovation, so much so that it is possible to define other distinctions regarding the phenomenon of innovation. Two important theorists, Henderson and Clark, in "Architectural Innovation: The

Reconfiguration of Existing Product Technologies and the Failure of Established Firms" (1990), analyze two particular binomials, the radical/incremental ones and the architectural/modular ones. Lastly, attention is paid to competence-enhancing and competence-destroying innovations.

- Radical and incremental innovations
- Architectural and modular innovations
- Competence enhancing and competence destroying innovations

Radical innovations have absolute novelty as their main feature. The change of course, in fact, must be significantly different from existing products and processes (Henderson and Clark, 1990).

Incremental innovations, on the other hand, represent the other side of the coin compared to radicals. They must not be unpublished and it is also possible to search for them even within the same company or sector and consist of marginal changes (Henderson and Clark, 1990).

Architectural innovations take the form of a change in the general structure of the system, which often occurs for company organization charts. These changes often affect the company project as a whole, also investing processes and products (Henderson and Clark, 1990).

Modular innovations, unlike architectural ones, foresee changes only in individual structural components, which do not involve substantial changes in the corporate project (Henderson and Clark, 1990).

Competence-enhancing innovations are a development of pre-existing knowledge. This evolution generates new models that incorporate previous ones (Shilling, 2000).

Competence-destroying innovations assume that new technology does not derive from inherent competencies or even render them inadequate (Shilling, 2000).

The Innovation Process in the Healthcare System

With specific reference to organizational innovation, Van de Ven (1986) defines it as the development and implementation of new ideas by people who engage in transactions with others over time within an institutional order.

From a sociological perspective, innovations are new (at least to the adopting community), making communication a necessary condition for adoption.

Organizational innovation can also be understood as the intentional introduction and application (within a group or organization) of ideas, processes, products or procedures, new to the relative unit of adoption, designed to significantly improve the individual, group, organization or broader society.

Therefore, the link between innovation and implementation is particularly crucial for the modernization of the current healthcare context. A healthcare innovation may refer to technology, a facility, an administrative system or a medical service that is relatively new to the industry as a whole and recently adopted by hospitals in a particular market area.

Healthcare innovations may be product-related, process-related or structure-related (Varkey, et al., 2008). The former may refer to goods or services, an example could be innovations in clinical procedures. The latter concern innovations in production processes or in the delivery and distribution of the product that allow for a significant increase in the value delivered to one or more stakeholders. Finally, structural innovations can have an impact inside and outside the company, creating new business models in the healthcare context.

Therefore, innovation is a complex and multidimensional process in any sector it is generated. However, individuals who decide to adopt innovation in the healthcare system go from a phase of knowledge of healthcare innovation to confirmation and then use of the innovation.

Knowledge and innovation represent the drivers of development for companies that operate in the current socio-economic contexts of reference. Rogers (1962) defines the innovation-decision-process as that process through which an individual (or other decision-making unit) goes from the first knowledge of an innovation to forming an attitude towards the innovation, to a decision to adopt or reject, to the implementation of the new idea, and to the confirmation of this decision. The five main phases of the process are listed below:

(1) knowledge,
(2) persuasion,
(3) decision,
(4) implementation,
(5) confirmation.

During this process, technology and communication have favored the development of clinical analyses by making available to stakeholders' modern methodologies that were difficult to access in the past (the remote transmission of images is one of the most striking cases). The increased dissemination of information has also enabled patients to carry out an innovation and development activity often favored by companies.

In this regard, Omachonu, and Einspruch (2010) define a healthcare innovation:

"as the introduction of a new concept, idea, service, process, or product aimed at improving treatment, diagnosis, education, outreach, prevention and research, and with the long-term goals of improving quality, safety, outcomes, efficiency and costs".

Any attempt to understand the innovation process in the healthcare system should start with an in-depth analysis of its challenges. According to Omachonu, and Einspruch (2010), five key stakeholders can be identified in the innovation process and each one must be represented through its needs, desires and expectations. They are reported as follows: doctors and caregivers, patients, organizations, innovative companies, and regulatory agencies. Studies show the ongoing difficulties in changing the behavior of physicians and health care organizations

As described above, the adoption of innovations is regulated by laws, which make the process slow for the necessary changes to be undertaken. As one might imagine, patients are the stakeholders who suffer the continuous inconveniences of this inflexible and unsuitable system for rapid market changes.

These difficulties lead to the adoption of attitudes that inhibit the diffusion of innovations and the learning of new organizational practices (Huntington et al., 2000).

To respond to such attitudes, new practices in patient care are traditionally carefully examined in their early development phase, so that potentially harmful innovations are not adopted but not even developed. Any attempt to model the process of health care innovation must take into account all five major stakeholders.

Therefore, the path of introducing new innovative technologies in the healthcare field must see a necessary synergy between research, industry and public institutions in charge of the NHS. This path still fails to connect the activities carried out by the three entities mentioned above. Very often their projects are disconnected from each other with a waste of energy and economic resources that could be channeled in the right direction.

Innovation in the healthcare sector continues to be the essential element and a driving force in the search for a balance between cost containment and quality of healthcare. In this restrictive vision of innovation, the most immediate thing that is taken into account is the cost of innovation represented by a new device or new drug (DiMasi et al., 2003). Until the 1990s, the pharmaceutical sector was conditioned by the Blockbuster Drug era, focused on basic and generic drugs with little propensity for innovation.

In a broader perspective, the costs of innovation must be evaluated in relation to their introduction, when, often, the considerable costs of equipment or instruments or even drugs are accompanied by structural changes (material and organizational) (Omachonu and Einspruch, 2010). Tangible costs are often not the only element of evaluation and the most current systems of evaluation of technological innovations tend to also evaluate intangible costs such as costs for structural adaptation, for staff training, for the establishment of complementary services such as help-desk or rehabilitation. The period of introduction of equipment or entry into the drug formulary must also be evaluated among the costs.

Therefore, this represents a strategic phase for company managers who must avoid costly premature introductions or, on the contrary, delay necessary insertions.

The diffusion of innovations in the healthcare system concerns a series of behaviors, routines and ways of working that are perceived as new by stakeholders, discontinuous with previous practices and aimed at improving health outcomes both in terms of administrative efficiency and in terms of costs or user (e.g. patient) experience (Greenhalgh et al., 2004).

Below, the author traces the innovations in the institutional and healthcare context in Italy and analyzes the diffusion path of new drugs with regard to classification, authorization, marketing, prices and related reimbursement systems.

DESIGNING THE FUTURE FOR SUSTAINING TECHNOLOGICAL CHANGE: FROM CLOSED INNOVATION TO OPEN INNOVATION

So far, attention has been paid to innovation and how it has spread over the years, also taking inspiration from various theories.

Understanding how the business model changes from before to after the diffusion of innovation is also necessary (Adams et al., 2006; Nambisan et al., 2017). For these reasons, to design the future of managerial approach to sustain technological change, firms are rethinking their business models from closed to open. Traditionally, two schools of thought refuted the focus on innovation. On the one hand, the social deterministic school argued that innovation resulted from a combination of external factors such as cultural and demographic changes and economic influences.

On the other hand, the individualistic school promoted the assumption that innovations resulted from individual talents who, through their innovative intelligence, experimented with new ideas. Over the past ten years, the literature on the "innovation" phenomenon has focused attention on two different models.

The first, called "market-based", argues that market conditions provide a context that facilitates or thwarts the scope of firms' innovation activity for innovation management measurement (Slater and Narver, 1994; Adams et al., 2006).

The key problem, of course, is that companies cannot always recognize opportunities in the market. Several theorists claim that few companies have the ability to effectively scan and search their environments for digital innovations (Nambisan et al., 2017).

The second model, the "resource-based" one, is believed that a market orientation does not provide a secure foundation for formulating innovation strategies for dynamic and volatile markets, rather a firm's own resources provide compliance more than major (Penrose, 2009). This model is based on the company and its resources, skills, and competences. It continues the assumption that when companies have precious, rare, and not easily emulated resources, they can frequently achieve a sustainable competitive advantage in the form of new innovative products.

In general, we can speak of innovation as a design and development process, in fact, until a few years ago, this concept was linked to something "closed", internal to the company that left no room for currents that came from external.

Chesbrough (2003), adopting a business strategy perspective, presents a persuasive argument according to which the innovation process has moved from a closed system, internal to the company, to a new mode of an open system that involves a series of actors distributed on the whole supply chain.

It is therefore possible to assume that this is the emphasis placed on the new knowledge-based economy that introduces the concept of open innovation (Chesbrough, 2003). The author just mentioned, is considered the founding father of the open economy model; in fact, he tries to explain that this concept can be understood as the antithesis of the traditional vertical integration model, in which R&D activities within a company led to products also developed internally and which were then distributed and marketed by the same company. This new model, on the other hand, systematically focuses on collaboration, ideas, and external resources with respect to the corporate perimeter traditionally understood.

From start-ups to the global market for ideas and patents, open innovation illustrates better than previous models why a company no longer needs to control, almost to own, the innovation processes from the beginning to the end of the supply chain.

It is inevitable that we cannot speak of a perfect model, which presents at least two critical points:

- it cannot be valid for all sectors: in fact, there may be sectors in which almost total control of the innovation processes is necessary and cannot be modified.
- open innovation implies a radical change of mentality: from controlling the innovation process we pass to influencing it, a step that is not at all obvious or simple that often does not allow companies to take this model into consideration.

Ultimately, the concept of "Closed Innovation" alludes to the behavior of a company that tries to measure itself with the market using its unique resources, internalizing all processes, and believing that they are the best means the way of being an 'agency.

This closed model has become obsolete due to increased worker mobility, better education, the growing presence of venture capital, ever shorter product life cycles, increasing competition, economic globalization, better use of information technology, and wide availability of knowledge from multiple sources (De Jong et al., 2008).

All these phenomena have prompted firms to consider another model of the economy that has as its pivotal points characteristics diametrically opposed to those of the model illustrated above.

The open innovation (OI) model is seen as the use of inbound and outbound knowledge flows that are converted into economic values. In this case, it is visible the presence of external partners who collaborate in the value chain by collectively developing ideas for a given company project. This phenomenon is called "crowd-sourcing". Diversified factors influence the open innovation process, as well as the intrinsic size of an organization.

Henry Chesbrough, believed to be the father of this current, says that "there are two important types of open innovation:

- out in;
- inside out.

The first allows organizational organizations to store inputs and contributions from outside, adapting them to their business models. It is this aspect of open innovation that has received the greatest attention, both in academic research and in practical industry.

Open inside-outside innovation allows organizations to transfer unused ideas outside to allow others to use them in their companies and business models. In this case, the model is less explored and therefore less understood, both in academic research and in industrial practice.

The concept of open innovation is in contrast with traditional closed innovation.

In past years, the idea was widespread that, in the field of Research and Development (R&D), product and process innovations should take place within the company itself in order to achieve competitive positions in the strategic business areas of reference.

This was because innovation was considered unique and successful if it came from the company's self-sufficiency. The development, communication, control, trade and distribution actions had to be carried out independently. "If you want something done well, you must do it yourself."

The closed innovation strategy involves hiring qualified employees who bring together their ideas, knowledge and skills, in order to find an advantageous solution for the company in which they work: innovative ideas are born within the company, and it is important not to share them with competitors.

As previously illustrated, it incorporates management, governance, skills and technologies, while also integrating the relationship of open innovation factors with other related organizations, institutions and agencies.

REFERENCES

Abernathy, W. J., & Utterback, J. M. (1978). Patterns of industrial innovation. *Technology Review*, *80*(7), 40–47.

Adams, R., Bessant, J., & Phelps, R. (2006). Innovation management measurement: A review. *International Journal of Management Reviews*, *8*(1), 21–47. DOI: 10.1111/j.1468-2370.2006.00119.x

Chesbrough, H. (2003). The logic of open innovation: Managing intellectual property. *California Management Review*, *45*(3), 33–58. DOI: 10.1177/000812560304500301

De Jong, A., Kabir, R., & Nguyen, T. T. (2008). Capital structure around the world: The roles of firm-and country-specific determinants. *Journal of Banking & Finance*, *32*(9), 1954–1969. DOI: 10.1016/j.jbankfin.2007.12.034

Dezi, L., Leone, D., Schiavone, F., & Simoni, M. (2019). The knowledge management of micro-firms in the crowd: Key challenges for successful operations. *Production Planning and Control*, *30*(10-12), 1005–1018. DOI: 10.1080/09537287.2019.1582105

Geissdoerfer, M., Vladimirova, D., & Evans, S. (2018). Sustainable business model innovation: A review. *Journal of Cleaner Production*, *198*, 401–416. DOI: 10.1016/j.jclepro.2018.06.240

Golinelli, G., & Dezi, L. (1997). *Reti, finanza, progetti*. Cedam.

Greenhalgh, T., Robert, G., Macfarlane, F., Bate, P., & Kyriakidou, O. (2004). Diffusion of innovations in service organizations: Systematic review and recommendations. *The Milbank Quarterly*, *82*(4), 581–629. DOI: 10.1111/j.0887-378X.2004.00325.x PMID: 15595944

Henderson, R. M., & Clark, K. B. (1990). Architectural innovation: The reconfiguration of existing product technologies and the failure of established firms. *Administrative Science Quarterly*, *35*(1), 9–30. DOI: 10.2307/2393549

Itami, H., & Roehl, T. W. (1991). *Mobilizing invisible assets*. Harvard University Press.

Leone, D., & Schiavone, F. (2019). Innovation and knowledge sharing in crowdfunding: How social dynamics affect project success. *Technology Analysis and Strategic Management*, *31*(7), 803–816. DOI: 10.1080/09537325.2018.1554858

Leone, D., Schiavone, F., Appio, F. P., & Chiao, B. (2021). How does artificial intelligence enable and enhance value co-creation in industrial markets? An exploratory case study in the healthcare ecosystem. *Journal of Business Research, 129*, 849–859. DOI: 10.1016/j.jbusres.2020.11.008

MacVaugh, J., & Schiavone, F. (2010). Limits to the diffusion of innovation: A literature review and integrative model. *European Journal of Innovation Management, 13*(2), 197–221. DOI: 10.1108/14601061011040258

Nambisan, S., Lyytinen, K., Majchrzak, A., & Song, M. (2017). Digital innovation management. *Management Information Systems Quarterly*, *41*(1), 223–238. DOI: 10.25300/MISQ/2017/41:1.03

Nonaka, L., Takeuchi, H., & Umemoto, K. (1996). A theory of organizational knowledge creation. *International Journal of Technology Management*, *11*(7-8), 833–845.

Omachonu, V. K., & Einspruch, N. G. (2010). Innovation in healthcare delivery systems: A conceptual framework. *The Innovation Journal*, *15*(1), 1–20.

Penrose, E. (1959). *The theory of the growth of the firm*. Blackwell.

Penrose, E. T. (2009). *The Theory of the Growth of the Firm*. Oxford university press.

Rogers, E. (1962). *Diffusion of Innovations*. Free Press.

Rogers, E. (2003). *Diffusion of Innovations* (5th ed.). Free Press.

Rogers, E. M. (2010). *Diffusion of innovations*. Simon and Schuster.

Schumpeter, J. A. (1934). *The Theory of Economic Development*. Harvard University Press Cambridge.

Selznick, P. (1957). *Leadership in administration: A sociological interpretation*.

Shilling, C. (2004). The body in culture, technology and society.

Slater, S. F., & Narver, J. C. (1994). Does competitive environment moderate the market orientation-performance relationship? *Journal of Marketing*, *58*(1), 46–55. DOI: 10.1177/002224299405800104

Teece, D. J. (2000). Strategies for managing knowledge assets: The role of firm structure and industrial context. *Long Range Planning*, *33*(1), 35–54. DOI: 10.1016/S0024-6301(99)00117-X

Van de Ven, A. H. (1986). Central problems in the management of innovation. *Management Science*, *32*(5), 590–607. DOI: 10.1287/mnsc.32.5.590

Varkey, P., Horne, A., & Bennet, K. E. (2008). Innovation in health care: A primer. *American Journal of Medical Quality*, *23*(5), 382–388. DOI: 10.1177/1062860608317695 PMID: 18820143

Chapter 2
Digital Transformation and Business Model Innovation

ABSTRACT

Digital transformation is a complex and heterogeneous process to which various professional, institutional, and societal actors contribute actively, thanks to access to innovative technologies. This chapter provides an overview of searching for strategic sustainability in digital transformation, focusing on its impact on business models and organizational capabilities. Considering the ever-evolving landscape of contemporary business, the role of business models, especially in the context of start-ups, has become increasingly crucial to sustain technological change. Noticing that the entrepreneurial ecosystem continues to evolve, along with the customer needs, the strategic use of business model innovation of these emerging enterprises is becoming more and more crucial. First, the chapter explores the main theories and different perspectives of business model innovation. Subsequently, the chapter examines how digital transformation drives business model innovation, introducing theories such as the Business Model Canvas and future perspectives in searching for sustainability.

INTRODUCTION

Digital transformation is the process of evolution of business models, based on technological progress, which involves significant changes. For optimizing new emerging technologies and their rapid expansion in human activities, the company must know how to reinvent itself, totally transforming all its models and processes. The transformation has adopted modern technologies by changing the corporate

DOI: 10.4018/979-8-3373-3466-0.ch002

culture, seeking to improve or replace existing approaches. Digital transformation is a cultural change that requires companies to get involved, experiment with new paths and models, operate, and offer value to customers in any sector. In fact, digitization has changed all economic sectors, from banking to commercials to the healthcare sector. Digital transformation is seen as a goal to be achieved. There may be various reasons why a company chooses to undertake it, but the prevailing reason is that it is essential, especially if you want to be able to move in an increasingly competitive market with continuous changes. Companies choose to adopt this transformation for various reasons such as to improve Customer Experience or how a company interacts with its customers at every stage of the purchasing process, or for increasing productivity or profitability. Technology is used as a fundamental tool for maintaining business continuity (Bozkus, 2023).

Digital transformation became essential during the covid-19 pandemic (Stalmachova et al., 2021); in fact, during this period measures were taken to limit the spread of contagion, especially by companies that have moved their operations and services online. There are clear advantages, for which it is important to adopt the transformation, such as:

1. Greater business efficiency.
2. Improved customer experience.
3. Increase in profits.
4. Alignment with the competition.

Thanks to it, processes and workflows are faster, easier and more efficient. This increased efficiency allows employees to focus on other key activities for the company and to complete projects and assigned tasks much faster. Digital transformation allows companies to get to know their customers more accurately, putting them first in their digital strategy to fully satisfy their needs and expectations. Today, users are always connected, informed and up to date and expect to find relevant and interesting information online based on what they are doing, where they are and the device they are using. This means that they:

- They will be more likely to try a new product.
- They will talk about the company with friends, relatives and acquaintances.
- They will buy from that company even if a competitor sells a product at a slightly lower price.

It is necessary for all companies to realize the benefits deriving from digital transformation and to adopt a competitive strategy in order not to lag behind their direct competitors. In recent years, many companies have embarked on the process of digitization by changing their culture and business organization, facing new challenges, interacting differently with customers and markets and using a different approach in managing their activities. It is therefore essential for companies to adopt digital transformation because it is a user need to increase efficiency, customer satisfaction and company revenues.

The fourth industrial revolution has totally changed people's habits. By now, almost unconsciously, one finds oneself connected with the rest of the world (Floridi, 2014). Precisely this change in habits at the same time entails a substantial change in the choices of products by consumers.

This process, that leads to the use of new technologies, is called digital transformation. One of the main characteristics of this revolution, however, is that digital transformation does not only concern the most powerful multinationals (such as the giants of Silicon Valley), but also SMEs or new start-ups. The latter, paradoxically, could redesign the current market in the event that large companies do not approach digital transformation correctly.

But how can companies change their structure and way of doing business without risking being cut off from the market in which they are inserted?

Capgemini (World leader in consulting and technology services), defines "3 digital capabilities" thanks to which companies can initiate this transformation:

1. *Customer Experience*: with the advent of social media, companies are able to indirectly receive the requests of their customers. Geographic location systems make it possible for businesses to know where their customers are at any time. Predicting purchasing behavior leads to a totally different sales approach on the part of companies, which by leveraging new digital technologies are able to promote an innovative shopping experience for the customer. The decline of the concept of Customer experience is undoubtedly Customer engagement, that is the emotional involvement between company and customer. Implementing a good customer engagement strategy is by no means trivial, since thanks to the personalization of the customer experience and the creation of both multichannel and omnichannel approaches, the company can improve customer loyalty, accentuating and strengthening the bond with it. The goal, therefore, is to try to shift attention from the product to the person with a customer centric approach. Several examples in different sectors are emerging in this case, from the use of e-commerce apps integrated with augmented reality technology that allow the vision of furnishing solutions or choice of clothing without having to physically go to the store.
2. *Operational Process:* to digitally transform a company, it is of primary importance to integrate new operational processes that aim to improve customer experience. The giants of Silicon Valley interpret new digital technologies as

matrices of opportunities to innovate and abandon those which until then were considered strengths and which are now obsolete.

3. *Business Model:* another digital capability is the introduction of new business models. Together with MIT (Massachusetts Institute of Technology), Capgemini identifies 5 main types of business models that lead to the digitalization of companies:
    1) The first aims to review sectoral standards, through the use of new electronic communication platforms (Car2go) that experiment with a new type of economy that takes the name of shared economy.
    2) This type, on the other hand, tends to introduce new platforms that replace existing products and services (Netflix, in the filmography sector).
    3) The third type concerns the creation of new digital products through the connection of customer data through wearable devices with smartphones and apps (for example Nike +).
    4) Other typology aims at revising delivery models, improving direct contacts with customers through Connected Car Services (such as UberEats or Glovo).
    5) The latter presupposes reconsidering the "performance bond", or the offer of flexible short-term insurance (such as pay-for-use for Mercedes Me navigation services).

Ultimately, it is deductible that companies cannot make use of all the tools made available by the new industrial revolution but must opt for those that are more functional for the entrepreneurial strategy, for the internal organization, for the sector in which it operates and for all the subjects that orbit around it (Lukito et al., 2022).

If, for example, when we talk about the second industrial revolution, the discovery of oil immediately comes to mind, the fourth industrial revolution does not have a single and revolutionary discovery as its daughter, but rather a bundle of enabling technologies that come to aggregate thanks to systemic paradigms. productive.

These paradigms are in continuous transformation and therefore they will involve innovations that are always different also on the sectoral level: of the process; organization; of product; of business model. Therefore, it is undoubted to say that the fourth industrial revolution has not reached the peak of its horizons but is on the contrary a revolution in the making. In the following sections, different enabling technologies made available to Industry 4.0 will be framed.

What Drives Organizations to Innovate and Digitalize Their Processes to Manage Technological Change

After making a general overview of the needs of digital capabilities in organizations, trying to explain as much as possible from what sources innovation derives, what forms and models are, and how it spreads, it would be appropriate to ask why and where companies should innovate to manage the technological change.

There are various reasons that push companies to do so and in addition, the benefits that can be drawn from it are identifiable regardless of the individual characteristics of the companies, such as the size of the company or the sector to which it belongs.

Innovation can in fact reduce production costs, increase product quality and open outlets on new markets.

In general, innovations that manage to provide cost or demand advantage over their competitors increase the company's profitability (Webster, 2004).

It is therefore completely normal to express the identified advantages and risks and obstacles that affect the innovative processes of companies.

The former can be divided between advantages and risks

- *Benefits:*

    1) Consolidation and strengthening of a position on the market;
    2) Differentiation from competitors;
    3) Increase in profits;
    4) Business growth;
    5) Increase in yields;
    6) Replacement of old products and services;
    7) Increase in the hypothetical life span of the company.

- *Risks:*

    1) Recurrent and high expenditure of costs and time;
    2) Failure of innovation attempts.

In innovating, companies, as previously mentioned, may encounter various obstacles, that can lead to significant negative consequences, limiting or even eliminating the implemented innovation project.

Some of these obstacles are present within the company and actually depend on it and are called internal brakes.

Those that instead originate from the outside are instead defined as external brakes.

- *Obstacles:*

    1) Lack of qualified personnel;
    2) Resistance to new technologies;
    3) Organizational problems;
    4) Funding for new projects.

- *Internal brakes:*

    1) High cost of projects;
    2) Amortization period of innovations too long;
    3) Poor aptitude of the company towards new technologies.

- *External brakes:*

    1) High risk of innovation being imitated by other companies;
    2) High risk compared to the innovation market prospects.

Innovation at the company level can concern products, processes, marketing activities and organizational issues. There is no doubt that, in a constantly evolving market like the current one, managing your business effectively and efficiently does not always bring the desired results. It is necessary to implement the process of acquiring knowledge that lasts over time. For this reason, we speak of technological innovation and strategic innovation, as in modern economies, the engine of economic growth has often been represented precisely by technology and business strategies adopted by companies to distinguish themselves from competitors (Arora et al., 2004). It is therefore possible to analyze the two topics in detail, subsequently highlighting the relationships that are created between them. To introduce strategic innovation within a company, it must be based on competitive positioning, which is useful for identifying where to place or relocate a company. To evaluate their competitive position, companies rely on the "Porter 5 Forces model", through which they are able to obtain more information regarding the details and intrinsic characteristics of the market in which they are placing themselves, thus being able to make effective

Figure 1. Porter 5 forces model (Adapted from Taha, 2018)

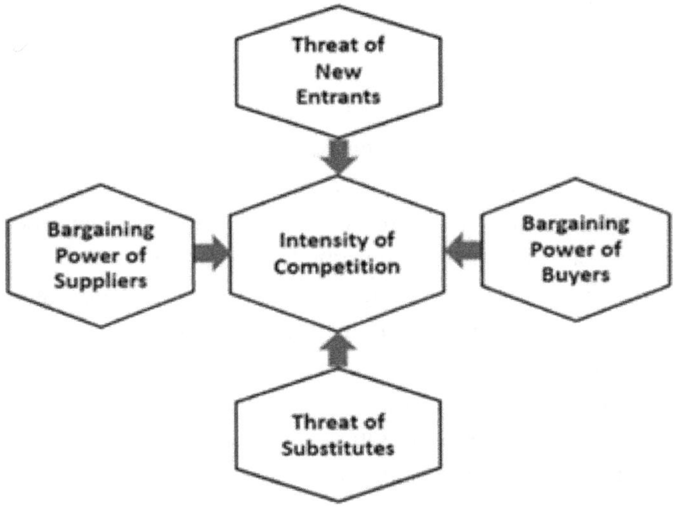

strategic decisions, and establish behaviors and attitudes to be adopted towards these forces

Therefore, after identifying a gap with its competitors, a company implements an innovative strategy that aims to modify this deficit and sometimes improve it by making what was previously a disadvantage, an advantage.

It is therefore possible to summarize this concept by taking up the now-famous sentence by Markides in "Strategic Innovation" 1999, which says:

A strategic position is simply the sum of the answers that a company gives to the questions: Who should I target as customers? What products or services should I offer them? How should I do this in an efficient way?

Strategic positioning is effective if the company is able to respond positively to these questions:

1. *WHO*: Who are the company's customers? By choosing a specific segment or a geographic market, it is possible to answer this first question.

Focusing makes it possible to better target innovative, productive and commercial energies at the service of a specific target. Truly understanding the target of the company's customers is the basis of good strategic positioning.

2. *WHAT:* What product or service does the company offer?

In this case, the company should redefine the characteristics of the products or services it offers on the market.

3. *HOW:* How to convey the offer?

Finally, by redefinition of HOW, we mean the ability to change the processes through which goods or services are conveyed to customers. It is possible to make this idea effective by redefining the distribution and communication processes with which the product or service is proposed to the target audience.

In summary, in order to carry out strategic innovation, companies must be able to choose these three fields of action in a distinctive way from other competitors, and they must also be equally creative in developing innovations in all three dimensions (Who, What and How) at the same time, since the development of an innovative strategy in one direction is doomed to failure.

Technological innovation can be understood as the application to the operation of the company, of advanced technical and scientific knowledge that makes available on the market versions characterized by a considerable functional improvement compared to previous versions, such as allowing gains in terms of opening new markets or expanding its market shares. Therefore, technological innovation is characterized by:

1. Improvement of technology until it changed
2. Meeting the needs of the ever-changing market
3. Opening up to new markets.

It is necessary to emphasize that there are real mechanisms that push the company to implement technological innovation. In fact, it can arise under the pressure of a technological push or demand-pull type process.

In technology push theories, the process of technological innovation arises from R&D: the latter would be independent of market events and would represent the main engine of innovative activity. According to this model, technological innovation would follow a linear process of an inverse type with respect to the demand-pull model, that is, from R&D to the market and not vice versa. The innovative activity could enjoy, at least initially, good autonomy with respect to market dynamics, moving towards those socio-economic theories that speak of products and goods not demanded by the market but, on the contrary, "imposed" by companies on consumers.

Scientists make partly unexpected discoveries; technologists apply them to the development of new products and designers turn them into prototypes to be submitted to the market. It is left to the production experts to find the most efficient ways to produce. Finally, marketing has the role of convincing potential buyers to buy the product

Demand-pull theories attribute a decisive role to the market in designing the path of technological innovation, having a concept diametrically opposed to technological pull theories. The innovative activity is fundamentally aimed at satisfying new needs that the market somehow manages to express.

Therefore, the most convinced theorists of demand-pull admit that in some way it is possible for companies to carry out a sort of recognition of any unsatisfied needs of consumers and as a result undertake an innovative research activity in order to satisfy them, through the improvement of products. existing or even creating new ones that better adapt to the wishes of the market.

However, an analysis of technological innovations is appropriate, for the taxonomy proposed by Upton (1996), who divides the innovations into:

1. *structural*: these concern the introduction or discontinuation of new implementations or even phenomena such as the outsourcing of corporate functions or complex assets. They modify the entire value chain.
2. *system*: they include a set of technical production solutions whose functionality is expressed in relation to the correct functioning of the software and not for management practices.
3. *philosophical or managerial*: the reason why this type of technological innovation has developed lies in the fact that even after having found solutions to a system, it is always possible to review the procedures and reorganize the activity with the related management processes.

Ultimately, it is possible to state that system solutions constitute the prerequisite for the growth and development of managerial approaches to the problems of the technical-productive function deriving from intense reciprocal interdependencies between system solutions and managerial solutions, which find in the former the substrate suitable for their use (Thompson, 1967).

The new innovative companies have an advantage because they are capable of implementing innovative processes to give rise to a virtuous circle of territorial development (Audretsch and Acs 1990). The latter states that 4 fundamental elements explain the advantages that new innovative companies have in the knowledge-based economy. These elements are innovation, employment, sector dynamics, and finally entrepreneurship.

The economic model that has been progressively affirming is also defined as a knowledge-based economy, to identify a development model in which knowledge becomes the means for creating value (Rullani, 2004).

In this scenario, software or new technologies are applied, which allow the management of the supply chain phases to benefit, by bearing sustainable costs and adopting eco-compatible operating procedures, thereby reducing waste of time, resources and energy through artificial intelligence and augmented reality. These are advanced technologies that enable processes and organizations to make rapid decisions.

Simply put, they offer greater transparency and flexibility, while simultaneously increasing the resilience of the entire system. In conclusion, the combination of advanced technologies such as software for predictive analysis and algorithms based on machine learning allows government bodies and companies to detect possible anomalies in processes early and make decisions in real time to protect costs and the environment.

Following the technological change, Artificial Intelligence is a discipline that combines the contribution of many sciences and offers entrepreneurs unlimited and infinite possibilities in the name of sustainability. In 2018, the European Commission identified several points to define AI:

- Fall within the scope of artificial intelligence these systems designed by man in the form of software (and hardware) that act in the physical or digital dimension and which, given a complex objective, perceive their environment through data acquisition, structured or not, interpreting them and reasoning on the knowledge or processing the information derived from them, deciding the best actions to take to achieve the given objective".
- "AI systems use logical rules or the assumption of a numerical model and can also adapt their behavior by analyzing the effects that their previous actions have had on the environment."

In this vein, AI includes different approaches and techniques, such as machine learning, reasoning and robotics. In short, we can say that artificial intelligence is the ability of a machine to demonstrate human capabilities such as reasoning, learning, planning and creativity. Augmented reality is a technology that allows the superposition of virtual content onto the real environment, creating a system that combines real-world objects and computer-generated objects and amplifies the sensory experience. Augmented reality is often confused with virtual reality: both are based on modifying, enriching, or replacing the physical environment with digital elements; the difference lies in the fact that augmented reality "enriches" physical reality without replacing it. Augmented Reality can bridge the gap between consumers, products, and product contents. The ability to overlay additional information, visual stimuli and interaction offers companies the ability to connect the digital world with the physical one in a targeted way. Numerous companies are starting to consider the potential of Augmented and Virtual Reality as a useful investment, also using it in new and innovative ways.

The next chapter offers an understanding of why it is necessary to design a new business model with knowledge, skills, and competitive abilities in mind, also by thinking about the technological change we are living in.

THE NEED FOR DESIGNING A BUSINESS MODEL

The term 'business model' is used to represent the fundamental aspects of a business, including strategies, purpose and organizational structures. It arises from the definition given by the economist and essayist Peter Drucker who in 1954 defined the business model as a tool capable of answering some questions relating to customers, costs and value propositions. The issues related to business models have been analyzed from different perspectives in multiple sectors, however, defining the boundaries of this phenomenon is not easy. In fact, there is no clear and universal definition to indicate business models, rather, there are numerous classifications, sometimes even incompatible (Zott and Amit, 2010).

Certainly, initially, the definition referred to the way in which companies used the great possibilities that the Internet and technology were starting to offer day after day. It is precisely for said reason that the early articles on this phenomenon date back to the 1990s.

One of the first and among the most important contributions of the literature on business models is the one provided by Timmers who addresses the issue of electronic commerce as a tool for creating business. Based on his research, we can identify a business model as an architecture for product, services, information flows or as a description of the potential benefits for the various players in the company or a description of the sources of profit. The author also establishes that the business model fails to provide contributions for the realization of the corporate mission, it is necessary that this is accompanied by a marketing strategy to understand what the functions and competitive advantages of a company are.

A business model can be defined as a business idea put into practice (Gassmann et al., 2013). The business idea consists of: the core strategy that defines the business objectives and the elements for which the offer differs; strategic resources relating to knowledge, skills and abilities; the value network characterized by the relationships between the company and suppliers who, with synergistic links, contribute to the creation of value by the company; the customer interface aimed at informing him about the characteristics, price and value of the product / service offered. The union between these four elements is provided by the so-called 'bridging elements': the customer benefits or the value brought by the business strategy through the customer interface; the configuration of the internal strategy which mainly concerns the combination of skills and processes; the boundaries of the company outlined by the value network dependent on the selected resources.

Business Model Canvas and its Elements

Another very detailed definition was provided by Osterwalder, Pigneus and Tucci who specify how a business model is a conceptual tool that contains a set of elements that through their relationship allow to express the business logic adopted by a company. Furthermore, it is the expression of the value that a company offers to one or more customer segments of the architecture of the company and its network of partners capable of creating, putting on the market and delivering this capital of value and relationship. to generate profitable and sustainable income streams. For the three authors, the business model is placed at the center of the 'Business Triangle' in which there are the strategy, organization and information systems of the company. External forces act on the three components such as the reference environment, standards, competitive forces and technological development that can modify the model in order to maintain its effectiveness.

For Teece, professor and organizational theorist, the business model articulates logic, data and others evidence that supports the value proposition for the customer and a viable structure of revenues and costs for the company that delivers that value. However, it is necessary that the business model is unique and inimitable by the competitors in order to maintain the competitive advantage of the company.

There are certainly some concepts that are re-proposed by the authors, including 'the value proposition' or the 'value offered' concerning the benefits that a company brings to its customer segment by providing the required products / services. The second element is the "profit model" or "profit formula" which, on the other hand, concerns the value that the company is able to obtain from the offer. It can be said, therefore, that what is important to define in a Business Model is on the one hand the value to be offered to the customer, on the other hand understanding how to derive the value for the company.

The configuration of a Business Model is a rather complex process that can be summarized in four main steps. The first of these concerns the identification of the critical success factors and the customer target. The study of these factors allows you to create an offer that can satisfy customers and that has an originality compared to the offer of competitors. The elements to be evaluated concern: the quality of the product / service, the breadth of the range, product availability, delivery times and price.

Once the critical success factors have been identified, it is possible to move on to the second step: identify which processes are best suited to be able to package an offer suitable for customers. Hence the customer value proposition which consists in identifying all the benefits that a company promises to offer in exchange for the price associated with the sale of the product / service. Once the customer value proposition has also been outlined, we move on to perfecting the Business Model. The third step concerns the identification of support processes such as human resource management processes, computerization and the definition of administrative and financial activities. Finally, the characteristics of the key resources of a business model can be defined. These resources are generally of three types: individuals who possess the resources necessary for the implementation of the model, technologies and financial resources. Once the configuration has been completed, the Business Model can be applied to the company.

A business model can be outlined by analyzing the nine basic building blocks that comprise the four main areas of a business: customers, supply, infrastructure and financial strength.

The Business Model Canvas was proposed by Alexander Osterwalder in his first work, Business Model Ontology (2004), and then developed by Osterwalder himself, Yves Pigneur and Alan Smith together with a community of 470 experts in 45 countries around the world. a strategic Business Design tool that uses visual language to create and develop innovative business models, thus allowing it to visually represent the way in which a company creates and distributes value for its customers. Today the Business Model Canvas is recognized as an international standard.

The elements arranged on the left of the model determine the efficiency of the organization, those on the right are the elements responsible for creating value.

At this point, we can see a certain similarity with the strategy defined 'Blue Ocean' defined in this way as, with this, the company " swims " in a blue ocean out of the red one of ruthless competition. The 'Blue Ocean' strategy is also based on the logic of reducing costs for the company and increasing value for the customer to completely neutralize the competition. The value proposition, which in the Business Model Canvas, occupies the central position represents the synthesis between these two objectives.

Customer Segments

The basic element relating to customer segments defines the groups of subjects to which the company wants to offer its product / service, that is, to those who intend to address their value offer.

We can identify customers as the beating heart of a business, as without customers no business could survive.

To be able to satisfy customers, the company needs to group into distinct segments based on homogeneous needs and behaviours. From this distinction, the company consciously chooses which segment to target, and on the basis of a deep understanding of the needs of each specific customer carefully designs a business model.

The different segments are built on the basis of the different needs that require and justify a different offer, the different distribution channels with which the groups of customers are reached, the different profitability, the dissimilarity with respect to the evaluation of the offer in relation to the price to be to pay.

There are different types of customer segments. By way of example, we illustrate the following:

- Mass market. In this case there is no real customer segmentation. The value proposition, the distribution channels, the relationships are quite standardized and focus on a large group with similar needs and preferences among the subjects.
- Niche market. It refers to a highly specialized customer segment in which the offer follows this specialization.
- Segmented market. There are segments of customers with similar needs but who need a degree of customization and differentiation of the final product. So, each segment is offered a slightly different value.
- Diversified markets. The company serves customer segments with completely different preferences, the model adopted by the company here seeks to offer value strongly based on the needs of the segments.
- Multi-sided markets. The company targets interdependent segments, in the sense that for the business model to work, the presence of all the segments involved is necessary.

The analysis of this 'building block' allows the company to answer the following questions:

Who are we offering value to?

Do all customers have the same importance?

Value Propositions

This element defines the set of products and services that produce value for customers.

The value proposition consists of all those benefits that the company is able to bring to the customer; it is precisely on the basis of this that customers choose one company over another. The value proposition is, in other words, the composition of products and services that the company produces to meet the needs and requirements of customers who belong to the segment to which the company is aimed.

The value can relate to quantitative elements such as price or qualitative elements such as design.

We identify some elements that can add value to customers:

- News. Some value propositions can satisfy new needs, never expressed before by customers, also based on the fact that there was no offer related to that need. For example, the production of smartphones represents a novelty as it has created a completely new industry in relation to the need for interaction and sharing of individuals.
- Performance. One way to create greater value is to increase the performance of a product. Using the example above, the smartphone industry has also focused on this: compared to previous years these are faster, enjoy more features and for these reasons the value offered is greater.
- Personalization. Each individual, according to their needs, is looking for products / services that can satisfy them, customization guarantees the customer to 'build' the product and the satisfaction of specific needs is greater.
- Brand / Status. Using and displaying products of a certain brand or category can represent value for the customer. For example, wearing a Gucci garment expresses wealth.
- Price. The value proposition can be represented by the adoption of a low price if the company is aimed at a price-sensitive customer segment. An example in this case can be represented by 'low cost' airlines which, thanks to the design of specific business models, are able to offer a service to customers at a low price.
- Accessibility. Another way of creating value is to increase accessibility to products that customers could not have available thanks to the use of new technologies and the adoption of innovative business models.

An in-depth study of this 'building block' allows us to answer the following questions:

'From what can the satisfaction of our customers derive?'

'What value can we transfer to customers?'

Channels

Channels represent the means through which the company communicates and transmits value to its customers. These have the task of helping customers evaluate the offer, raise awareness of the product / service they are going to purchase and provide after-sales assistance to buyers.

The channels are characterized by five phases:

- Awareness. This concerns the way in which the company makes the public aware of what it offers.
- Evaluation. It concerns the help that the company can provide to customers with respect to the evaluation of the offer.
- Purchase. The company establishes the methods by which the purchase of products / services can take place.
- Distribution. Identifies the way in which the offer reaches the customer.
- After sales. It refers to the support offered to customers after the purchase act.

Each channel can include all or part of these phases. Furthermore, the channels are divided into direct or indirect and proprietary channels or partner channels. It is the organization that will choose whether to use its own channels or partner channels or both, to reach customers.

The proprietary channels can be direct such as, for example, the website, and indirect such as the stores managed by the organization. The partner channels, on the other hand, are indirect and include different distribution methods: wholesale, web or retail. The partner channels also allow the company to obtain a wider range of action even if they generate lower profit margins. Those owners lead to a greater profit but greater expenses are required for activation and their proper functioning. Therefore, proper channel management occurs if the company manages to combine the two methods in order to create customer satisfaction and maximize revenues.

The analysis of this basic element allows us to answer the following questions:

'How do customers want to be reached?'

'What are the most convenient channels that integrate customer needs?'

Relations With Customers

The basic element 'customer relations' describes the different types of relationships between the company and the customer segment to which it refers. The choice of relationships to maintain depends on the reasons that push the company to establish them. These can arise from the desire to acquire customers, to retain them or from the simple reason to increase sales.

We can define different types of relationships ranging from personal to automated ones.

- Personal assistance. This interaction is based on personal relationships. A company employee interfaces directly with the customer when he needs assistance at the time of the sale or after it has been completed.
- Dedicated personal assistance. This type of relationship is deeper and more specific. This means that a specific employee is dedicated to the individual customer. This type of assistance occurs, for example, in private banking services, where dedicated financial consultants are at the service and follow high-value clients.
- Self-service. The company does not entertain any type of relationship with the customer but limits itself to providing the necessary information so that the customer can find a solution for himself.
- Community. Many companies create online communities to foster knowledge exchange and to solve mutual problems. This increases customer engagement real and potential ones, which in any case can acquire information by accessing the community.
- Co-creation. In this way, the customer is directly involved in the creation of business value: through active participation, he makes choices that can change the value proposition.

A careful evaluation of this element allows you to answer these kinds of questions:
'What kind of relationship do customers expect to establish?'
'How do the relationships adopted integrate with the rest of the business model?'

Revenue Flows

The revenue streams represent the money that the company obtains through the sale of products / services. To secure one or more revenue streams, the company should understand for what value a customer segment is really willing to pay a certain price.

Revenue flows can be characterized by different price definition mechanisms such as fixed price list, bargaining, auction sales or market-dependent prices.

In a business model we can identify different revenue streams deriving from two payment methods:

- Revenues resulting from payments in a single solution.

- Revenue from recurring payments.

There are also various methods capable of generating revenue streams.

- Sale of goods. It is the best-known and most used source of revenue. The income derives from the sale of the ownership of goods that may belong to different genres and categories.
- Usage fee. The revenue stream is generated thanks to the use of a service. The more the service is used, the more revenue streams increase.
- Registration fees. In this case it is the sale of access to the service that generates revenues, as is the case with the subscription fees to the Spotify platform, access to the music library takes place through the monthly payment of a fee.
- Loan / Hire / Lease / Rent. Revenue flows derive from the sale of the right to use an asset for a specific period of time. The advantage of those who provide these services lies in the fact of receiving recurring revenues. Renters or tenants enjoy the advantage of incurring expenses for a specified period of time without having to bear the costs related to the ownership of the property.
- Licenses. Revenues are generated by allowing customers to use intellectual property in exchange for a license right while still retaining the property.
- Advertising. Revenue streams are generated by expenses incurred to advertise a product or service.

As for prices, they can be defined in different ways. In the case of fixed prices, the company will establish them, for example, on the basis of the volume of business and the customer segment; in the case of dynamic prices, they will be defined in relation to real-time market trends, negotiations with partners and other variables.

The 'building block' of revenue flows allows the company to meet the following questions:

'What value are customers willing to pay for?

'How does the revenue stream improve overall revenue?'

Key Resources

The key resource lockout establishes the assets and strategic assets needed for the business model to function. Each activity is developed through key resources that may be owned by the company, rented or acquired from partners and are divided into:

- Physical resources. They include tangible assets such as plants, buildings, machinery and everything needed for the production of goods and services. Physical resources are essential for carrying out business, think of the

Carrefour group which owns a huge global network of stores, without these the business could not take place.

- Intellectual resources. They include know-how, patents and trademarks. Developing intellectual assets turns out to be quite difficult but once created and owned they can offer substantial value and are increasingly important in setting up an effective business model.
- *Human resources.* These are fundamental in a business model, especially in the service sector. Facebook wouldn't exist without its programmers like Ikea without its designers for developing creative ideas.
- *Financial resources.* They include particular financial resources such as lines of credit, cash or a set of stock options that allow the company to hire important employees or secure supplies.

The study of this element allows us to answer the following question:
'What resources are needed for the value proposition?'

Key Activities

The basic element relating to key activities describes the most important tasks that the company must perform to operate successfully.

Like key resources, they are needed to support value propositions, reach customers, maintain relationships with them, and generate revenue. The identification of said activities depends on the chosen business model.

The key activities can be classified as follows:

- Production. These activities involve the design, creation and distribution of high quantity products. Manufacturing firms focus on this type of activity in their business models.
- Problem solving. They concern the identification of new solutions for customer problems. Organizations dominated by problem solving activities are typically those that offer services such as firms that provide consulting services.
- Platform / network. This activity requires that the business model adopted by the company identifies the platform as a key resource. Companies like eBay or Facebook need platforms, these are essential for the functioning of the business.

The analysis of this 'building block' allows the company to answer the following questions:

'What key activities increase the value proposition?'
'Which are best suited to our distribution channels, our relationships and revenue streams?'

Key Partnerships

This basic element defines the network of suppliers and partners necessary for the functioning of the business model. The existence of partners in a company is very important; they create alliances to reduce costs, increase resources and optimize their business models.

The distinction between the different types of partnerships can be made as follows:

- Strategic alliances between non-competing companies: this is the case with suppliers or companies that are located within a single production chain, for which there is a link but not of a competitive nature.
- Strategic alliances between competitors: in this case there is the redistribution of risks and costs across multiple entities.
- Joint Venture: consists of an agreement between companies that could generate new business models as a result of their collaboration.

The reasons that push companies to create partnerships are:

- Optimization and economies of scale. It is possible that the company does not have all the necessary resources available or that it does not carry out all the activities internally. Partnerships for the optimization and implementation of economies of scale usually allow the company to reduce costs and often involve outsourcing or sharing of infrastructures.
- Reduction of risk and uncertainty. It often happens that companies form a strategic alliance in a sector, despite being in competition with each other. In this case, partnerships can be useful for reducing the risk and uncertainty associated with the competitive environment.
- Acquisition of particular resources and activities. As we have already pointed out, there are few companies that manage to possess all the key resources and are able to carry out all the activities necessary for production and its support. For this reason, the role of partnerships is fundamental: thanks to these relationships, the company can provide itself with particular resources and can carry out certain activities.

The study of the 'Key Partnership' element allows the company to resolve the following issues:

'Who do we identify as our key partners?'
'What benefits do we get from these?'

Cost Structure

This last element defines all the costs that the company must incur to make the business model work.

In the mapping of the Business Model Canvas, the cost structure is left as the last element. This is because it derives from: the creation and distribution of value, from the maintenance of relationships with customers, from the generation of revenues which are all activities that involve costs.

The business model can be structured while also managing to keep costs low, as happens with low-cost airlines.

It may be useful to distinguish the cost structures of business models into two classes: cost-based and value-based.

- Cost-based. Cost-based business models aim to reduce costs in any area possible. Each reduced cost represents an additional value to the product you intend to offer.
- Value-based. Some companies focus more on creating value than on cost reduction. For these, it is important to create products with precious raw materials through exclusive processing, or to convey to customers a very high value of the good / service.

Furthermore, we can analyze the characteristics of the cost structures:

- Fixed costs. They are those that remain unchanged regardless of the volume of goods or services.
- Variable costs. They are those that undergo variations in relation to the production volume.
- Economies of scale. They allow for a reduction in costs following the increase in production volume.
- Economies of scope. Cost reduction by increasing the range of action.

Thanks to the study of the cost structure, the company can consciously answer these questions:

'What are the most important costs in the model?'
'What is more expensive?'

The next chapter offers an understanding of how the concept of sustainability could be intended, also by thinking about strategic sustainability in business models.

SEARCHING FOR STRATEGIC SUSTAINABILITY IN BUSINESS MODELS INNOVATION: DIFFERENT PERSPECTIVES TO MANAGE TECHNOLOGICAL CHANGE

The organization must evaluate the importance of sustainability in developing its business models to achieve a competitive advantage. The definition of sustainable development can be traced back to the Brundtland Report, 1987, entitled "Our common future" in which it identifies it as the ability to meet the needs of current generations without compromising the ability of future generations to achieve their own.

Over time, the concept has undergone extensive evolutions such as extending towards a global vision that qualifies sustainability as synonymous with well-being and progress, identifying it in the ESG (Environmental, Social and Governance) criteria divided as follows:

- Environmental responsibility implies the protection of the environment and its systems. The production process implies the reduction of natural resources such as energy and water and limits waste and refuse by focusing on recycling. Pollution and greenhouse gas emissions are mitigated by reducing the use of chemical substances and opting for the use of raw materials obtained from one's own territory.
- Economic responsibility, based on a fair and long-term vision of economic development and well-being;
- Social responsibility contributes to the improvement of living conditions and the well-being of people. Currently, efforts are being made to improve the working conditions and remuneration of workers employed in textile production. The economic and business environment is constantly evolving: globalization, innovation and sustainability are factors that affect the evolution of companies operating in various sectors.

In recent years, new eco-friendly consumer trends have increased the demand for sustainable products, making sustainability a competitive advantage to focus on to increase turnover and attract potential customers. According to Nielsen, a leading market research company, 52% of buyers are more likely to purchase a product resulting from the production process of a company that adopts sustainability policies, highlighting society's growing attention to the ethical behavior of the company.

In terms of a sustainable approach, there are two types of companies: pioneers and newcomers. The pioneers, representing 10% of the total, anticipated the times with visionary strategies, and today they are reaping the fruits of their actions, such as cost reduction, process efficiency, and improved reputation. The "newcomers", on the other hand, have recently started their journey, when sustainability was already considered a significant value, and now they continue in the wake of customer and stakeholder expectations.

A company is considered sustainable when it avoids the creation of negative impacts resulting from its own actions and those of its subsidiaries, identifying and preventing any external damage by exercising due diligence. Among the recent initiatives of 2022, the Commission has encouraged responsible and sustainable corporate behavior along the entire supply chain by issuing the Due Diligence Directive, aimed in particular at large companies with a net turnover of more than 150 million euros and more than 500 employees.

The degree of corporate sustainability can be assessed with reference to national and international criteria identified by the Global Reporting Initiative (GRI), an international non-profit organization created for the purpose of developing reports on the sustainable performance of any entity. The result of the GRI standards is a table divided into four series in which, in addition to the ESG factors, it is possible to determine corporate sustainability from a universal perspective. The total estimate is included in the non-financial statement, a mandatory report for specific companies in order to encourage attention to sustainability. The publication of sustainability reports not only allows the public to know the company's commitment to sustainable development but also allows for the financing of business projects through access to tenders and subsidized financing. Thanks to these advantages, 69% of Italian companies have launched a corporate sustainability project in recent years, of which 44% have outlined a strategic plan consisting of quantitative objectives; the remaining 15% of Italian companies say they plan to have a sustainable development plan in the future: this is what emerges from the EY study "Seize the Change Sustainable Futures", carried out on a sample of over 300 Italian companies operating in different sectors. To support Italian entrepreneurship, the National Recovery and Resilience Plan (PNRR) has allocated 37% of its funds to the implementation of sustainable corporate policies.

New Theories and Models Towards Strategic Sustainability

Business model innovation is a fundamental factor in improving the performance of a company. However, it is important to ask where this innovation should be directed in order for it to be effective and bring the desired results. Surely one of the most important innovative drivers at the moment is represented by the theme of sustainability.

In this paragraph, the book tries to illustrate the path to developing a sustainable business model, identifying the characteristics that distinguish it. Subsequently, in the last two paragraphs, the theme of sustainable business model innovation will be treated and an overview of innovation in small and medium-sized enterprises will be given. Initially, we will try to give a definition to the concept of sustainable business model, then we will try to understand why a company should have an interest in using it, and finally, we will carry out a study to understand the link between sustainability and performance. As for the concepts described above, there is no unequivocally accepted definition in the literature for the sustainable business model, so it is useful to analyze the main thoughts of scholars in order to extrapolate a definition that can be used in the continuation of this analysis. The common features of the sustainable business model that involves many projects are:

- The sustainable business model modifies some characteristics and objectives related to the traditional business model.
- Principles, concepts or objectives related to the theme of sustainability are incorporated within the business model.
- Sustainability is built into value proposition, value creation, or in the value capture mechanisms. Based on these common features, it is possible to identify the sustainable business model as a business model in which it is incorporated: a pro-active perspective towards stakeholder management, the simultaneous creation of monetary (economic value) and non-monetary (social and environment) for a vast pool of stakeholders (internal and external to the company) and which adopts a perspective not oriented only to short-term logics, typically economic, but which is capable of adopting a long-term perspective (Geissdoerfer, Vladimirova, & Evans, 2018); time horizon in which the progress of one's policies at an environmental and social level can be appreciated as well as the positive repercussion of these two aspects in the economic dimension.

After having understood in a general way the main characteristics related to the sustainable business model, the most interesting definitions contained in the literature will be analyzed below. Porter and Kramer (2011) argue that the sustainable business model is a source of competitive advantage in which value proposition, value creation and value capturing mechanisms incorporate the principles of sustainability, this brings economic benefits to the companies that adopt it. Boons and Lüdeke -Freund (2013) identify four characteristics that distinguish a sustainable business model from a traditional business model:

- The value proposition integrates environmental and social value with economic value. Sustainable supply chain management is introduced.

- The sustainable business model must create relationships with consumers and other stakeholders in order to develop more sustainable production and consumption systems.
- In designing the financial model, both economic benefits and the company's social and environmental impacts must be considered. Abdelkafi and Täuscher (2016) believe that the sustainable business model provides value not only to the firm's consumers but also to the environment and society. Geissdoerfer et al. (2018) understand a sustainable business model as a set of elements. The interrelationship of these elements and the interaction with stakeholders creates, delivers, captures and exchanges sustainable value to many stakeholders. Finally, the study by Roome & Louche (2016) is very interesting. It integrates the components of a traditional business model identified above. In the business model, the value process creation is central and directly linked to value proposition, in the sustainable business model in addition to the value creation value is taken into consideration destruction.

In the traditional business model, the creation of economic value involves a simultaneous destruction of value for other actors (stakeholders) now or in the future (Lovins, Lovins, & Hawken, 1999). Companies that decide to implement a sustainable business model must therefore worry, in addition to the creation of value for themselves and for society, also to keep the value that is destroyed under control, trying to minimize as much as possible the negative impact of their activity (Roome & Louche, 2016).

For this analysis, the definition of sustainable business model used will be: "A business model that integrates within its fundamental components (value proposition, creation, capture) the sustainable logic, and which adds to these components the case of destroyed value".

After understanding the definition of sustainable business model, we will try to understand why companies should adopt it and the advantages of its application. As the report states from the Brundtland Commission (1987): "Sustainable development calls for the integration of environmental and social issues into the decisions that shape economic and social development whether they are made by the public or private sector." In the light of this statement, it is necessary to ask why companies should pose the problem of introducing a logic of sustainability within their business models. The theme of sustainability concerns the entire planet, and as it is now known, the situation at an environmental and social level worsens over time, above all due to exponential phenomena such as the demographic increase, which in turn causes the birth of other sub-phenomena, also with an exponential movement.

It is clear that a global problem of such importance and scope does not directly concern only public organizations and non-profit organizations but involves every person and organization on the planet. Organizations become fundamental players in this reasoning: they are made up of a set of subjects who, through the coordination of their actions, aim at achieving the objectives they set for themselves. Generally, the main objective that organizations have dealt with over time is the achievement of a profit, which is typically achieved through the use of materials that are transformed and resold. The link between sustainability and companies is therefore clear, companies represent and manage the productive resources of the economy (Bansal, 2002).

According to Roome & Louche (2016) transformation arises from the need for companies to remain constantly competitive, this is achieved while contributing to a more sustainable development. Companies therefore continue to focus on their primary purpose, i.e. the production of economic value, but rather broaden this perspective of value creation by also extending it to the creation of social and environmental value, considering in this process a wider pool of stakeholders from satisfy. To build a truly sustainable business model, these three values must be pursued simultaneously, without one of the three prevailing over the others. To do this, the company necessarily collides with paradoxes, just think of the difficulty of reconciling a better salary for employees with the achievement of better performance at an economic level.

To successfully achieve sustainability within the firm these paradoxes must be accepted, rather than eliminated, and it lies in the ability of a firm's managers to simultaneously manage seemingly conflicting issues (Hahn, et al., 2015). Over the years, attention to sustainability has increased sharply. By now practically all the large companies globally have decided to tackle this issue, with different methods and purposes, making sustainable dynamics almost a trend. Although all efforts to improve environmental and social conditions are important, it is nevertheless necessary to distinguish real changes from a sustainable perspective from phenomena that concern marginal changes in production structures.

A paradox refers to a situation in which seemingly opposite elements coexist (Lewis, 2000), these elements appear to be acceptable individually but together seem irreconcilable or incompatible.ù

Eco-design and eco- efficiency improvements have allowed for a reduction in energy use, the number of resources used, emissions and waste. However, these changes have proved ineffective in countering the growing use of resources globally and consequently have not substantially benefited the reduction of environmental impacts (Bocken, Short, & Evans, 2014). As Schaltegger (2012) states, one of the main challenges is the design of business models that allow the company to acquire economic value for itself by offering social and environmental benefits. While quality and efficiency improvements have a direct link to profit creation (reducing waste is generally associated with an economic benefit), it is not clear how the provision of social and environmental benefits can be translated into an increase in the firm's profits or competitive advantage (Bocken, Short, & Evans, 2014).

To conclude the study on the sustainable business model it is necessary to extend the analysis also outside the company itself. For years the logic of competition has been dominant, and it was believed that the fastest way to competitive advantage was to fight with competitors. For this reason, the competitive pressure has been growing more and more, above all due to the phenomenon of globalization which has brought this competition beyond national borders.

The logic of sustainability aims to change this paradigm, introducing the concept of collaboration between stakeholders. Value is no longer created by each firm individually but by bringing together many organizations through informal agreements or formal alliances (Beattie & Smith, 2013). Through the collaboration of many companies, it is possible to carry out large-scale initiatives and implement changes from a sustainable perspective that would have been unattainable individually. The focus shifts from achieving competitive success to creating social, environmental and economic value for all, in order to improve the global living conditions of current and future generations.

DIFFERENT POSSIBILITIES FOR INNOVATING YOUR BUSINESS MODEL

In the current evolutionary scenario, companies must necessarily adopt appropriate strategic solutions and manage the transformation of the organizational structure correctly. The transformation and innovation of business models allows you to enable new business capabilities identified as necessary, activate new value paths, innovate the role and position in the value chain, create more flexible and lean organizations, reduce fixed corporate costs freeing up financial resources and creating new business organizations capable of continually reconfiguring themselves. It is also essential to review the approach to the market by developing products based on customer needs, taking care of the image of the company and the quality of the products / services offered. It is important to know how to exploit technology to increase the information to be processed and reduce uncertainty.

The winning companies will be those capable of combining innovation, relationships of collaboration and partnerships with customers and suppliers, sharing of knowledge, critical and continuous analysis of both one's business and that of competitors, aptitude for developing strategies and planning activities, maximum attention to human resources with the introduction of development and incentive plans to attract and retain talent.

Some emblematic examples of companies that have exploited technology and that have gone far beyond the simple concept of innovation are Apple, Starbucks and Amazon.

In 2001, music players, for example, were already on the market but Apple was the first company to include music distribution among its activities thanks to its great innovative capacity with the launch of the iPod.

Amazon, by exploiting its assets and distribution channels, has managed to enter the e-reader market without having any skills in the electronic device sector. In 2007, Amazon became the leader in the e-reader industry.

Figure 2. Business model innovation as a subset of business model design and re-configuration (Adapted from Geissdoerfer, Vladimirova, & Evans, 2018)

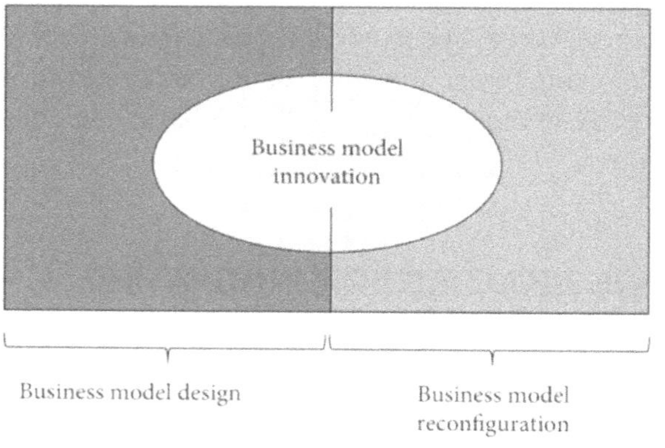

Innovation can also take place in widely established sectors, in which the products traded are commodities. The case that best expresses this possibility is that of Starbucks; the company has innovated by selling a common product for which the customer is willing to pay a higher price for the experience that, together with the product, is offered. So, more than on the product as such, innovation concerns everything that revolves around it: the creation of a comfortable environment, customer service, the ability to access Wi-Fi (Saric and Mikolasik, 2021).

If the product resulting from the innovation process were included in an equally innovative business model, then the company would be a winner; business model innovation becomes a priority for companies in the growth phase.

Think of the story of Blockbuster and Netflix; the young firm leveraged existing technology to innovate a business model on which Blockbuster had built its success. The novelty brought by Netflix concerns the possibility for the customer to choose films and TV series avoiding moving to the store. Blockbuster underestimated the threat of the start-up's innovative business model, and, in 2010, the company went bankrupt.

At this point we can identify the good reasons for innovation.

- Aim to be a leader, not a follower! The first reason refers to being first in a 'Red Ocean', that is, in the competitive market. Innovation is indispensable; it concerns the introduction of novelties in the process or a product that guarantees greater results or benefits than competitors. It gives companies the opportunity to come out of a red ocean, to discover a Blue Ocean, free from competitors.
- Swim in a fish-free sea. It refers to the fact of considering oneself alone within the Blue Ocean: discovering new customers, creating new needs, going as far as no one else has been able to do it. To do this, a strong capacity for innovation is required, with the appropriate tools and models that allow the development of undiscovered possibilities.
- Choose the most effective innovation. Not all innovations have the same impact on the success of the company, so it is necessary to choose the most effective one. Numerous studies have shown that the most effective innovation is the one concerned with the business models. This is because the innovation of the individual areas, especially that relate to the product, is easily replicable by the competition, while the company that manages to innovate every single function makes the difference much more solid. All this happens thanks to the use of the Business Model Canvas.

At this point we can say that Business Model Innovation (BMI) is a form of innovation based on the exploration and exploitation of new business opportunities.

The application of this implies a change of mentality and innovation becomes a continuous process capable of questioning the current business model.

The different possibilities for innovating your business model can be classified into:

- Industry Model Innovation: consists of innovating the value chain by moving towards new industries, redefining existing ones or creating new ones.
- Revenue Model Innovation: innovation, in this case, concerns the identification of new ways to generate revenues. This can be done by changing the configuration of the product-service value mix or through new pricing models.
- Enterprise Model Innovation: changing the role played by the company within the value chain can imply changes in relations with employees, suppliers, customers and the configuration of assets and skills.

- Open Innovation: it is an innovation system in which companies look outside the company boundaries, also exploiting any sources of external ideas to advance their business.
- Collaborative Entrepreneurship: is based on the creation of value that arises from the development of new ideas resulting from the sharing of information and knowledge.
- Open Business Model: companies open their business models by actively seeking and exploiting ideas from the outside and allowing unusual internal technologies to flow outwards, where other companies can unlock their economic potential.

The process of starting up or renewing the business requires on the one hand creative skills for the identification of new managerial opportunities for the implementation of a new or renewed business model as well as interpersonal and communication skills to involve the people of the company in the change process. It is necessary to create business models that satisfy market needs that have not yet been met; introduce new technologies, new products or new services; improve existing markets; create new markets. All this is possible with the use of a tool: the Business Model Canvas.

CONCLUSION

The innovation of business models is a fundamental step for companies operating in a scenario in which new economies such as Sharing and Crowd are born and evolve: it is necessary to found business models capable of grasping the new characteristics of these economies by evaluating and analyzing the new way in which the creation of value takes place, the management of resources and production capacities and the methods of accessing resources.

The birth of new economies has brought with it the need to redefine the elements of the organization. The novelty was in fact radical: new actors and existing actors with new roles, new types of interactions, new types of products and services. At this point, it is necessary to define an innovative business model (Amini and Jahanbakhsh Javid, 2024).

REFERENCES

Abdelkafi, N., & Täuscher, K. (2016). Business models for sustainability from a system dynamics perspective. *Organization & Environment*, *29*(1), 74–96. DOI: 10.1177/1086026615592930

Acs, Z. J., & Audretsch, D. B. (1990). *Innovation and small firms*. MIT press. DOI: 10.1007/978-94-015-7854-7

Ali, A., Davis, E., Martinez, C., & Brown, L. (2024). The Role of Organizational Culture in Driving Innovation and Competitive Advantage. *International Journal of Management. Business Economics (Cleveland, Ohio)*, *1*(1).

Amini, M., & Jahanbakhsh Javid, N. (2023). A multi-perspective framework established on diffusion of innovation (DOI) theory and technology, organization and environment (TOE) framework toward supply chain management system based on cloud computing technology for small and medium enterprises. *Organization and Environment (TOE) Framework Toward Supply Chain Management System Based on Cloud Computing Technology for Small and Medium Enterprises (January 2023).International Journal of Information Technology and Innovation Adoption*, *11*, 1217–1234.

Arora, A., Fosfuri, A., & Gambardella, A. (2004). *Markets for technology: The economics of innovation and corporate strategy*. MIT press.

Azeem, M., Ahmed, M., Haider, S., & Sajjad, M. (2021). Expanding competitive advantage through organizational culture, knowledge sharing and organizational innovation. *Technology in Society*, *66*, 101635. DOI: 10.1016/j.techsoc.2021.101635

Beattie, V., & Smith, S. J. (2013). Value creation and business models: Refocusing the intellectual capital debate. *The British Accounting Review*, *45*(4), 243–254. DOI: 10.1016/j.bar.2013.06.001

Bocken, N., Boons, F., & Baldassarre, B. (2019). Sustainable business model experimentation by understanding ecologies of business models. *Journal of Cleaner Production*, *208*, 1498–1512. DOI: 10.1016/j.jclepro.2018.10.159

Bocken, N. M., Short, S. W., Rana, P., & Evans, S. (2014). A literature and practice review to develop sustainable business model archetypes. *Journal of Cleaner Production*, *65*, 42–56. DOI: 10.1016/j.jclepro.2013.11.039

Bozkus, K. (2023). Organizational culture change and technology: Navigating the digital transformation. In *Organizational Culture-Cultural Change and Technology*. IntechOpen.

Burton, I. (1987). Report on reports: Our common future: The world commission on environment and development. *Environment, 29*(5), 25–29. DOI: 10.1080/00139157.1987.9928891

Floridi, L. (2014). *The fourth revolution: How the infosphere is reshaping human reality*. OUP Oxford.

Francisco, C. (2024). The Power of Influence: Leading by Example for Greater Role and Productivity. *Available atSSRN* 4862247.

Gassmann, O., Frankenberger, K., & Csik, M. (2013). Revolutionizing the business model. In *Management of the fuzzy front end of innovation* (pp. 89–97). Springer International Publishing.

Geissdoerfer, M., Vladimirova, D., & Evans, S. (2018). Sustainable business model innovation: A review. *Journal of Cleaner Production, 198*, 401–416. DOI: 10.1016/j.jclepro.2018.06.240

Hahn, T., Pinkse, J., Preuss, L., & Figge, F. (2015). Tensions in corporate sustainability: Towards an integrative framework. *Journal of Business Ethics, 127*(2), 297–316. DOI: 10.1007/s10551-014-2047-5

Lewis, M. W. (2000). Exploring paradox: Toward a more comprehensive guide. *Academy of Management Review, 25*(4), 760–776. DOI: 10.2307/259204

Lukito, D., Suharnomo, S., & Perdhana, M. S. (2022). Transformation management capabilities for digital transformation initiatives: A construct conceptualization in alignment with the dynamic capabilities framework. *Journal of Organizational Management Studies, 1*, 1–22. DOI: 10.5171/2022.845443

Naveed, R. T., Alhaidan, H., Al Halbusi, H., & Al-Swidi, A. K. (2022). Do organizations really evolve? The critical link between organizational culture and organizational innovation toward organizational effectiveness: Pivotal role of organizational resistance. *Journal of Innovation & Knowledge, 7*(2), 100178. DOI: 10.1016/j.jik.2022.100178

Nosratabadi, S., Mosavi, A., Shamshirband, S., Zavadskas, E. K., Rakotonirainy, A., & Chau, K. W. (2019). Sustainable business models: A review. *Sustainability (Basel), 11*(6), 1663. DOI: 10.3390/su11061663

Osterwalder, A. (2004). *The business model ontology a proposition in a design science approach* (Doctoral dissertation, Université de Lausanne, Faculté des hautes études commerciales).

Osterwalder, A., Pigneur, Y., & Tucci, C. L. (2005). Clarifying business models: Origins, present, and future of the concept. *Communications of the Association for Information Systems, 16*(1), 1. DOI: 10.17705/1CAIS.01601

Rashid, A., Asif, F. M., Krajnik, P., & Nicolescu, C. M. (2013). Resource conservative manufacturing: An essential change in business and technology paradigm for sustainable manufacturing. *Journal of Cleaner Production, 57*, 166–177. DOI: 10.1016/j.jclepro.2013.06.012

Roome, N., & Louche, C. (2016). Journeying toward business models for sustainability: A conceptual model found inside the black box of organisational transformation. *Organization & Environment, 29*(1), 11–35. DOI: 10.1177/1086026615595084

Roome, N., & Louche, C. (2016). Journeying toward business models for sustainability: A conceptual model found inside the black box of organisational transformation. *Organization & Environment, 29*(1), 11–35. DOI: 10.1177/1086026615595084

Rullani, E. (2004). La fabbrica dell'immateriale. Produrre valore con la conoscenza.

Saric, D., & Mikolasik, M. (2021). The impact of electronic services on traditional services. *Developments in Information & Knowledge Management for Business Applications, 2*, 305–334. DOI: 10.1007/978-3-030-76632-0_11

Stalmachova, K., Chinoracky, R., & Strenitzerova, M. (2021). Changes in business models caused by digital transformation and the COVID-19 pandemic and possibilities of their measurement—Case study. *Sustainability (Basel), 14*(1), 127. DOI: 10.3390/su14010127

Taha, T. M. A. M. (2018). Competitive Analysis of the Global Oil and Gas Industry using Porters Five Forces Model.

Teece, D. J. (2000). Strategies for managing knowledge assets: the role of firm structure and industrial context. *Long range planning, 33*(1), 35-54.Itami, H., & Roehl, T. W. (1991). *Mobilizing invisible assets*. Harvard University Press.

Thompson, R. F. (1967). Foundations of physiological psychology.

Upton, M. (1996). *The economics of tropical farming systems*. Cambridge University Press. DOI: 10.1017/CBO9781139172868

Webster, E. (2004). Firms' decisions to innovate and innovation routines. *Economics of Innovation and New Technology, 13*(8), 733–745. DOI: 10.1080/10438590410001686824

Zott, C., & Amit, R. (2010). Business model design: An activity system perspective. *Long Range Planning, 43*(2-3), 216–226. DOI: 10.1016/j.lrp.2009.07.004

Section 2
Value Co-Creation in Digital and Sustainable Economies

Chapter 3
Managing Value Co-Creation in Sharing Economy and Circular Economy

ABSTRACT

This chapter discusses theories and values co-creation aspects of the sharing economy and circular economy applied in the digitalized era. The sharing economy has deep origins in the history of humanity but is finding new life recently to overcome some modern obstacles. What can be said, however, is what is fundamental in all definitions of the sharing economy, namely the need to have three parts. The first is the one who wants to make underutilized asset proofs available to achieve value co-creation. The first is the one who wants to provide their underutilized assets (resource owner) to a platform (sharing platform) which has the task of acting as a mediator for those who need to use these assets (resource user). The chapter also offers an overview of the circular economy, a new business model born to turn waste into resources, often applied in supply chain management. Additionally, thanks to selected case studies, the chapter will provide practical insights related to the theory.

DOI: 10.4018/979-8-3373-3466-0.ch003

INTRODUCTION

Developing Innovation Towards Sustainability: Historical Phases and Triple Helix Model

Understanding where innovation comes from is a particularly difficult task for the different sources from which it can arise. They depend on the type of company and the sector in which the latter is inserted. Furthermore, it is possible to define two types of sources of innovation: internal and external.

The "internal sources" are outlined by the Research and Development (R&D) activity, the creativity, and the innovative spirit of the company's employees. The characteristic of creativity is undoubtedly fundamental for a company aimed at innovation.

In fact, it can depend on the knowledge, and skills, or know-how of individuals within the company. It was an objective for scholars to find a method that encouraged the creativity of the company's internal actors.

One of the most famous is the "suggestion box" conceived by Patterson in 1895, but still, focus groups, brainstorming activities, and training courses, are useful to encourage management to introduce dialogue tools that intersect the skills and knowledge of all within the firm (Boer et al., 2017).

Of equal importance is the research and development activity, managed by subjects capable of generating products and processes that meet the needs of the company.

The "external sources" are represented by figures capable of finding the best way for innovation:

- public research bodies, such as universities, hospitals, and research centers;
- public funds.
- non-profit organizations.
- private organizations and foundations.

Therefore, innovation can also arise from the interactions and relationships that are established with other actors with different backgrounds and positioned at different levels, thus creating a complex system of sources, where innovation can be the result of an integration between several sources Over the years, various theories have been developed regarding the combination of firm size - innovative capacity (Acs et al., 2005).

In general, it is possible to list four historical phases, each characterized by the emergence of new and more effective types of innovation players (Lazonick, 2007):

- the pioneering company;

- the great Fordist enterprise.
- the small-medium district enterprise.
- the new research-based company.

All reflect and highlight the characteristics of a specific historical, economic, and social moment. It is therefore possible to define as many models as possible of innovation generation.

In the phase that takes the name of "Proto-Entrepreneurial", the emblem is the figure of the inventor, who researches experimental techniques to create prototypes that in many cases have represented steps forward for the technology of all humanity (Marquesan et al., 2024).

The continuous development of the blue-collar and white-collar class has made it possible to promote mass markets, which have introduced the so-called "Fordist" large enterprise onto the economic scene, which has based its competitive advantage on large volumes of standardized production and at competitive costs.

The prevailing impression, also refuted by scholars such as Schumpeter in 1942, and subsequently, Chandler and Galbrait, is that big business could be understood as the real engine of innovation.

By modifying the levels of specialization and flexibility, the Italian districts and local production systems made up of networks of small and medium-sized enterprises have been able to provide an effective response to the complexity and changeability of tastes and personalization needs expressed by the market.

Through the division of labor between companies linked by collaborative relationships, it has been possible to obtain economies of variety and flexibility that the large Fordist enterprise by its nature aimed at centralization and standardization was not able to achieve (Becattini, 1999; Rullani, 1999). In recent years, the affirmation of a definable technology-based model has been evident, which is increasingly based on the ability to innovate, or rather on the ability to produce new scientific knowledge. For this reason, the development of technology transfer mechanisms from the research system to the business world is becoming increasingly important.

To be precise, so far, we have tried to analyze the sources of innovation analytically, paying particular attention to the combination of internal and external sources. At this point, it is important to understand the public-private relationship in the innovation scenario.

Specifically, the involvement of industries, public administrations, and universities is fundamental. Furthermore, the latter is indispensable for creating an economically competitive environment, being centers of knowledge and promoters of technology transfer for companies. Towards the beginning of the new millennium, a model centered on this system of relations was tested, which can create a favorable context for the transfer of knowledge and consequent development of innovation, by creating a form of knowledge.

Figure 1. Triple helix model (Adapted from Leydesdorff, 2012)

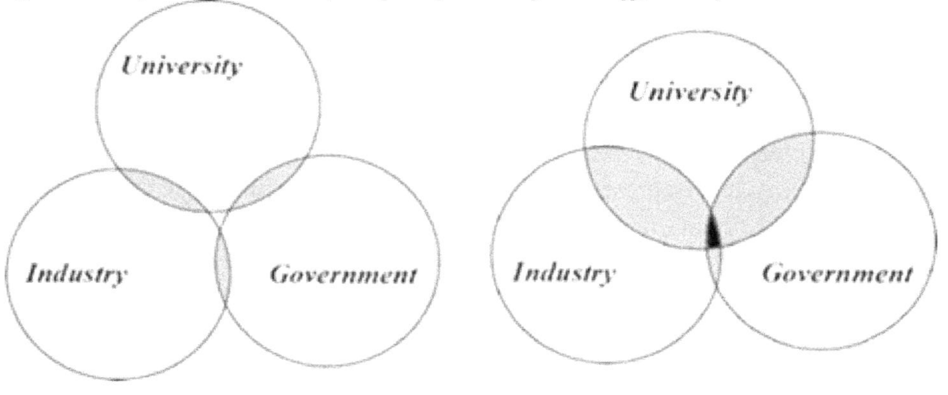

The creators of this theory that takes the name of the "Triple Helix Model" are Etzkowitz and Leydersdorff, who affirm that universities, the private sector and public administrations all operate on the same level, in a joint model, which manages to bring out structures hybrids that produce innovation thanks to the interdependence of the different needs, the different knowledge and skills possessed by the three propellers that make up the model. Furthermore, according to Etzkowitz, one of the key aspects of this theory concerns the relationship between the Triple Helix and the National Innovation System (SNI) introduced by the OECD to first conduct national reviews of research. systems, and then extended to innovation systems to consider the greater complexity of the innovation system, of which the research system is only one component.

Although the subjects are the same, the system promoted by Etzkowitz is totally different from that adopted by the SNI, since if in the latter each sphere operates following a single axis taking into consideration the traditional vision, in the Triple Helix, the subjects operate according to two axes: an X-axis in which each of them plays its traditional role, and a Y-axis in which each subject takes on new roles. This integration of roles, which, however, must not question the main one of each sphere is the main axiom of the "Etzkowitzian" theory (Leydesdorff and Etzkowitz, 1996).

In this sense, Open Innovation, the new innovation model promoted by Chesbrough discussed above, can be interpreted as the formalization of industry changes following the spread of new digital technologies. The Triple Helix and Open Innovation, therefore, are complementary models that follow the development of the economic and social sphere.

As already reiterated in the first lines of this paragraph, universities play a particularly important role in the development of knowledge, since alongside the two fundamental objectives of training and research, universities pursue a third mission, that of favoring direct application, enhancement and use of knowledge to contribute to the social, cultural and economic development of the company.

It is inevitable that over time, the link between research and companies has become more and more entrenched given the tendency of companies to use the research infrastructures of universities for their R&D objectives, thus indirectly transferring part of their costs to the state.

As often happens in all research sectors, also in this case the triple helix model has been the subject of discussion by various theorists, who have defined this model as too limited in terms of the number of stakeholders involved. Given that, as previously repeated, innovation is understood as the engine of the current economic system, it took a short time before we immediately started talking about a new model that in some respects surpassed the one experienced by Etzkowitz. This model takes the name of "Quadruple Helix".

It is therefore possible to deduce that a fourth is added to the three helices explained up to now. This intuition is based on the assumption that this model, unlike the previous one, strengthens the importance of companies and organizations operating in the field of creativity, thus becoming the fourth helix.

The authors of this model, Carayannis and Campbell, place as main characteristics, on the one hand, the potential of culture, and on the other, the influence of society on national and local innovation systems (Carayannis and Campbell, 2009). years of the new millennium, the quadruple helix was the model that most lent itself to being taken into consideration when it came to innovative development. Previously a clear distinction was made between open and closed innovation, in fact, in recent years this new way of innovating (open) has brought with it changes also regarding the models to be taken into consideration.

Today it becomes practically impossible not to consider the intervention of civil society in the macro-area of innovation, thus risking the creation of a further model, that of the "Quintuple Helix". The so-called "Collaborative City" or CO-City is based on a three-level urban governance model: sharing, collaboration and polycentricity. The integration between the different actors and local knowledge are the characteristics to start from in an innovative perspective also at the local level (Iaione, 2016).

Apart from some obvious differences, observing this table shows that the concepts reported share many characteristics. However, it can be deduced that the work carried out by Etzowitz was fundamental and preparatory for the formulation and development of the models that followed.

On the other hand, to respect the direction in which the firms must design their paths, managers have to consider aspects as social and economic resources. Organizations have for many years prioritized the management of economic resources to the detriment of the latter two, causing many problems that the world still faces today. This is because economic development not linked to environmental and social logics is useless and, moreover, ends up having negative repercussions on the economic sector itself. However, the interest of companies in sustainability is not confined only to a question of good conduct and social responsibility.

In recent years, many studies have been undertaken which have highlighted how the introduction of sustainable logic in a company can be seen as a source of competitive advantage and a tool to increase one's performance. It is therefore very interesting to explore the link between sustainability and performance, by analyzing the main considerations set out in the literature. Much research argues that if companies want to contribute to more sustainable development, they need to rethink and reshape their business models (Roome & Louche, 2016; Bocken, Short, & Evans, 2014). Internationalization has made global competition more complex, to keep pace with the market it is necessary to innovate as the traditional concept of business model struggles to find appropriate solutions (Nosratabadi, et al., 2019). One of the most interesting innovative forms to pursue is the introduction of sustainability within one's business model. Thus, the sustainable business model can replace the traditional business model concept as sustainable competitive advantage has surpassed traditional competitive advantage. Initially the concept of sustainability was introduced in order to make companies more environmentally and socially responsible (Rashid, et al., 2013), today instead the sustainable business model can be used as a source to obtain a competitive advantage (Porter & Kramer, 2011).

A New Dominant Logic: From Product to Service

The studies conducted by Vargo and Lusch (2004) have shown how, between the end of the 20th and the beginning of the 21st century, a new dominant logic in entrepreneurial marketing was established. In particular, before the focus was on tangible resources and transactions, in the aforementioned period new perspectives emerged that highlighted the importance of intangible resources and the "hidden" value inherent in the creation of new relationships. For Vargo and Lusch (2004), this evolution involved the transition from Goods-Dominant Logic (G-D Logic) to Service-Dominant Logic (S-D Logic). Since the beginning of the twentieth century, the logic of commercialization was based on the Goods-Dominant Logic, that is, on the exchange of tangible goods from a first seller to a second buyer.

The material consistency, therefore, presented itself as a fundamental requirement as it allowed the buyer to easily identify and focus on the value associated with it, leaving out some aspects of the exchange including, for example, the relationships, the subjects involved, and the methods.

The perspective focused on tangible good was based on these aspects:

- One of the objectives of economic activity was to find resources, produce, distribute and sell a tangible asset.
- To be sold, these material goods needed a certain value and a certain utility that could be transmitted to the consumer. This value, incorporated in the asset, had to be higher than that offered by the competitors to be purchased by the customer.
- The decisions taken by the company had to be made with a view to the economy: try to maximize its profit from the sale.
- It was necessary to achieve efficiency in the production process. To do this, the company continued with the standardization of the product and process.

Subsequently, thanks to numerous theories that have established a strong relationship between product and customer experience, the centrality of the product is gradually replaced by the figure of the customer, by the emotions and inner experiences that he manages to live in the buying process, from its involvement.

In the face of changes in society and consumers, Goods-Dominant Logic has had to leave room for more modern logics such as Service-Dominant Logic, capable of responding to new emerging needs and since providing a service to consumers obviously cannot to qualify as a standardized activity, this new perspective (service-dominant logic) requires a continuous learning process (Mascarenhas et al., 2006).

Services now pervade all economic activities, involving any production system and any organization. The emerging importance of services with respect to goods, traditionally understood, and the determining role of the same in all economic transactions has stimulated scholars, professionals, and business experts to engage in the search for models, paradigms, and theoretical constructs capable of describing the new value generation processes more effectively. From this point of view, Service-Dominant Logic represents a change of perspective with respect to traditional paradigms, capable of overturning the existing link between goods and services, revisiting the considerations related to their exchange and consequent use.

Service-Dominant Logic defines service as " an application of skills through actions, processes and performance aimed at producing a benefit for themselves and for third parties, directly or indirectly connected " (Vargo, Lusch, 2004a). According to this perspective, the set of resources of a company, consisting of specialized skills and adequate knowledge, represent fundamental elements for success and survival. According to this logic, the goods no longer represent the only object of the transaction, rather they appear as an instrumental element for the provision of services, considered as the real protagonists of the interactions and agreements between the parties to an exchange.

The main aspects of the new Service-Dominant Logic perspective theorized by Vargo and Lusch (2004) are:

- The identification and development of core competencies, or knowledge and skills that represent a possible competitive advantage for the company.
- The identification of other subjects such as "potential consumers" who can benefit from these skills.
- The development of relationships that involve consumers in the creation of value propositions that are able to meet their specific needs.

Service-Dominant Logic requires an evolution from logic focused only on The product. In fact, it is necessary to collaborate, be flexible, able to learn and to adapt to consumer needs. A system focused on service rather than product, implies that value is co-created with the consumer rather than being confined only to the tangibility of the product. The studies of Haeckel (1999) lead to highlight how the success of a company can be found in the abandonment of a perspective "I produce and

I sell "(which we find within the goods-dominant logic) in favor, on the other hand, the adoption of a strategic perspective that places the consumer at the center of all the activities. Furthermore, the studies by Vargo and Lusch (2004) highlight how the consumer is not only interested in the purchase of a tangible good but how he is led to purchase "resources" that can lead to obtaining service if properly combined.

In their definition, resources are understood as "skills and abilities intended to generate a service and a profit for both the consumer and the seller" (Vargo and Lusch, 2004).

The analysis conducted focuses on two main types of resources: the "operand resources" and "Operant resources" (Vargo and Lusch, 2004). Operand resources are those tangible resources on which a form of transformation can also be carried out. Operant resources, on the other hand, are given by the combination of intangibility and invisibility. These are dynamic resources and not static like operand resources. Operant resources are linked to people, to information, organizational dynamics and relationships that are created. In the ransition from Goods-Dominant Logic to Service-Dominant Logic the most important resources, i.e. those on which companies must try to base their competitive advantage, are those that fall within the classification of operant resources.

In the transition from Goods-Dominant Logic to Service-Dominant Logic, greater importance is also placed on creating value for the consumer. In this regard, yes they can identify two different types of value: "exchange value" and "value of use ". The exchange value is linked to the Goods-Dominant Logic while the use value concerns the Service-Dominant Logic.

In Goods-Dominant Logic, the value was created and produced by the company and later transferred to the market to customers, these two subjects, companies and customers, were clearly distinguished.

The point of view of Service-Dominant Logic is different in which, instead, producers and consumers are no longer conceived as two distinct entities but collaborate and cooperate in the joint creation of value. In fact, we speak of the co-creation of value. All the features of Service-Dominant Logic are now an integral part of the new economy companies that are facing new business models that are increasingly appreciated and valued in the current competitive scenario.

MANAGING VALUE CO-CREATION IN NEW ECONOMIES

The various actors involved in entrepreneurial activity control the creation of knowledge value; this identity, transform and reproduce the acquired know-how from time to time. The sharing of know-how constitutes, in a complex economy, a factor capable of generating a competitive advantage for the company.

The value of knowledge can be expressed in various types of activities such as the production of goods and services, the capacity for customer satisfaction and loyalty, and the anticipation and understanding of changes in the external environment. Therefore, it is necessary that managers, thanks to their ability to manage knowledge, have a clear vision of the strategic actions to be taken, evaluate the objectives in relation to the resources possessed, the interactions with the environment, and managerial skills. We can therefore identify the intangible resources on which the economic performance of the company depends, as advantages for the competitiveness of the company.

Knowledge, as underlined, has now assumed a fundamental role for strategic business planning: know-how, for example, has always been the basis of business development, and is considered the starting point for the realization of new ideas. and business innovation. However, the knowledge-based economy must keep pace with the evolution of society and the market. The current internal company context and the external environmental one present numerous differences compared to the past, such as to produce an impact on the business models adopted by companies, oriented towards technological development, with a flexible organization with a high knowledge content. Wanting to summarize the considerations made, the element that inevitably affects the structure and implementation of the strategies is knowledge and its sharing. In fact, every company, whether it follows a more traditional approach or follows business models typical of the new economy, always acts on two levels: a material one, relating to assets and an intangible one, which concerns the implementation of knowledge management.

The goal of creating value is linked to the need for growth and development of the company and reflects the long-term effects of corporate strategies regardless of their size or sector to which they belong. This goal fuels pro-active behaviour and the focus on a result, furthermore, the creation of value can only be a goal shared by all the players in the company, whether they are shareholders, managers or employees.

In managing a company, activating cooperation channels between internal and external elements allows you to find solutions that no one could achieve alone, this applies both to the transmission of information and knowledge and to the creation of goods and services.

Traditional companies were closed in the make or buy alternative: do something yourself or, if you can't, buy something already done to be able to resell it to your customers. Innovative companies, in search of a competitive advantage, are formed by social networks of producers, consumers, stakeholders, which activate a continuous flow of knowledge, experiences and to create new goods / services in co-creation.

The term co-creation refers precisely to companies looking for innovative ideas among consumers; they introduce business strategies that focus on the realization of a corporate value shared with the customer; to the vision of the market as a place where companies, consumers, employees, partners, suppliers, share resources, combine them and renew them to create value through new forms of interaction.

Cooperation is made possible by the value proposition, the objective, the purpose of what the company proposes and if this proposal succeeds in reaching and aggregating individual collaborators or groups, these produce value for themselves and for the company.

The companies that will be able to seize this opportunity will have the "ability to overcome their opponents in achieving their primary objective: profitability" (Grant, 1999). In other words, they will obtain what can be defined as a competitive advantage. To be able to create a competitive advantage, the company will have to focus on product or process innovation, and to do so it must carefully know its composition and inter-organizational relationships with suppliers with whom it creates stable relationships over time. In practical terms, it is a question of understanding which activities to focus on in order to be more competitive.

The Sharing Economy and the Different Types

In the current global landscape, we are witnessing an ever more accentuated diffusion of new consumption models based on sharing and collaboration that allow individuals to exchange and share goods, spaces and skills, promoting lifestyles that prefer saving, the redistribution of money. and socialization. We thus pass from possession to access, from purchase to reuse, from ownership of an asset to its simple use. Sharing practices are now becoming a dominant trend: there is more and more talk of Sharing Economy. The research by PricewaterhouseCoopers, an international network that provides management and strategic consulting services, identifies the sharing economy as "an emerging ecosystem that monetizes under-utilized production capacities, or favors the loan, rent, fragmentation of micro-skills. ". According to the American magazine " Time " it is among the ten ideas that will change the world. In fact, among the advantages we find those of an environmental nature such as the reduction of waste and the environmental and economic impact including savings, earnings, facilitation and the possibility of doing business and socially through sharing, socialization and community creation.

The phenomenon, which started in the United States around 2008 and established itself in Europe and Italy around 2013, is now applied to various economic and social sectors and declined in various ways, in the current context in which the development of new technologies and the awareness of a necessary better exploitation of resources has created the need to review the ways of production, use and consumption of the same. Although the sharing of goods has always been a mutual habit between people who are close to each other, in recent years the concept of sharing has evolved from a community practice to a profitable business model. In the absence of a single definition that can bring scholars and experts on the subject to agreement, we provide distinctions and definitions that allow us to focus on the characteristics and facets of the phenomenon.

Reporting what was stated by the steering committee of "Sharexpo - Milan shared city for Expo 2015", the term Sharing Economy can be translated into two different terms, which indicate the main events:

- Collaboration that occurs when several people network with the aim of creating a project from which everyone will benefit, even individually;
- Sharing that occurs when a group of people pool resources for the production of goods or services useful to their entire community.

For the European Commission's Business Innovation Observatory, the Sharing Economy refers only to "those companies that develop business models based on accessibility for peer-to-peer markets and their user communities". These are those companies "whose value proposition consists in the creation of a match between a peer who owns a certain resource (goods or skills) and a peer that needs that resource, within the required time frame and against reasonable transaction costs ".

According to the theory of Rachel Botsman, co-author with Roo Rogers of the bestseller 'What's Mine is Yours. The Rise of Collaborative Consumption ', the expressions Collaborative Economy, Collaborative Consumption, Peer Economy, Sharing Economy are frequently used as synonyms, but more precisely:

- The collaborative economy is the large whole that encompasses the whole: "an economy based on networks, distributed and in turn formed by interconnected communities and individuals, which transforms the ways in which we produce, consume, finance and learn". Production, consumption, methods of financing and education are the four areas that the collaborative economy hosts in its universe.
- Collaborative consumption is one of the four key components of the collaborative economy. It can be defined as "an economic model based on the sharing, exchange, trade, or rental of goods and services that privileges access over property" and that "is redefining not only what we consume but also how". It includes three different phenomena: the creation of new markets for unused goods; collaborative lifestyles; "product service" systems (paying for temporary access to the asset rather than buying ownership).
- The Sharing Economy is also a specific subset of the collaborative economy, in which underutilized resources, from physical spaces to professional skills, are shared by some users for a monetary or purely personal benefit, allowing them to be used more efficiently.
- The Peer Economy, in the same way, is part of the collaborative economy, being identified by "person-to-person markets based on mutual trust, which facilitate the sharing and direct exchange of goods or services". If on the one hand it perfectly overlaps the "Peer2Peer" concept of the Sharing Economy, on the other it also includes a slice of the sector related to Collaborative Production, referring to virtual markets dedicated to "do it yourself".

The four terms do not describe distinct phenomena; on the contrary they indicate different shades of the same macro phenomenon that is revolutionizing the way we buy and understand traditional economic relations and on which our economic system has traditionally been based. In fact, we can see that the economic activities that fall within the sharing economy are in fact nothing new: sharing, bartering, exchanging, renting have always been forms of economic relationships, they have always been an integral part of the economy, and they are still part of it today. The big news of the sharing economy is the scale in which sharing takes place and the means used. Sharing traditionally took place only between a few people, towards whom relations of knowledge and trust were often nourished through close agreements in person. However, sharing takes place on a much larger scale, involving strangers living miles away and using digital technologies.

To better understand this phenomenon, it is necessary to identify the factors that contributed to the birth and development of this new economy. Many believe that collaborative economies arose as a natural response to the financial crisis of the turn of the century, but also as a desire for a more sustainable lifestyle that puts the individual and the environment more at the centre, and as a response to a growing need to feel part of something in a world that is increasingly vast, placing at the centre of the business logic no longer the product offered but the satisfaction and involvement of the customer in the use of an innovative service.

Among the social factors that have contributed to the development of the Sharing Economy we can better analyze:

- The change in the system of values. It is important to note how strong individualism and personal well-being have left more and more space for completely different values such as those of belonging and sociality. Consequently, we are witnessing the birth of a new approach to consumption; the satisfaction of needs is generated, no longer only with the purchase of products, but from the search for consumer experiences that can strengthen the new system of values. Furthermore, the consumer is increasingly sensitive and attentive to environmental, ethical and sustainability issues; he attributes positive values to companies and organizations that act in compliance with these values and tends to favor them in the choice and purchase process.
- The increase of the world population. This phenomenon will favor the exchange generated by the Sharing Economy, as supply and demand will grow and it will be easier to bring these two factors together in a sharing environment.
- Desire for communication. Individuals have a constant desire to communicate, socialize and create communities. The sharing economy model is aimed at a new desire to communicate, both through social networks and in everyday life, feeling part of a community.
- Factors of sustainability. Consumer practices have a great impact on the environment, for this reason, sharing makes it possible to increase the responsibility of companies which, through the implementation of new exchanges, recognize the seriousness of the problem.
- Generational altruism. A social norm taught by the sharing economy is to believe in a community that works and collaborates together in order to ensure a sustainable future for future generations.

Explaining the birth of the Sharing Economy only with social factors is not enough: the author of " The Mesh " L. Gansky states that "business mesh models are flourishing thanks to the growth of social media, the Internet, wireless networks and mobile devices. "

In fact, the advent of the Internet and Web 2.0 since the early 2000s has favored the development of new technologies which have allowed users to interact with each other, exchange information, dialogue with any other user in any country of the world in real time. In a second moment, the birth of Social Networks favored peer to peer exchange, favoring the meeting between supply and demand. Technology has brought a great wave of innovation also within the payment system: payment by credit card and fast billing help customers to have confidence in the online purchasing operations that are the basis of many Sharing Economy platforms. . The relationship between money and individuals is less and less physical and more and more virtual: in this way it will be possible to speed up transactions between those who buy and those who sell.

From the point of view of economic factors, those that have moved the consumption models from the traditional ones to those of the new economy are:

- Economic crisis of 2008. The crisis of 2008 was characterized by both a liquidity confidence crisis. Initially, linked only to the local US financial sector, the crisis became systemic and global in scope, and involved not only the economic system but also the social and value system. In this context, the model of the Sharing Economy has proposed itself as a valid alternative to the traditional system, as an opportunity for a rebirth using a more sustainable and equitable approach in the face of the evident weaknesses of our economic system being questioned. The Sharing Economy proposes a completely different value system: it prefers the principle of access and exchange to that of ownership and purchase, service to the product, trust, collaboration in competition and the short supply chain with the elimination of all intermediaries and direct contact between buyer and seller. A further central concept is empowering, i.e. all collaborative consumption services and platforms have the consumer as their main focus: they are conceived and created by consumers and have the purpose of serving the latter. The model that imposes itself is therefore a model with horizontal forces, power and control. The traditional hierarchical, policy enforcement and top-down control system is abolished.
- Monetize unused resources. Resources that have a longer life cycle are no longer used for a limited period of time and then thrown away: in the perspective of the Sharing Economy, resources are shared and made available, creating added value, avoiding waste and making use efficient of the same.

- Increased financial flexibility. Globalization and communication have made it possible to have new earning opportunities: owners will no longer earn on the sale of the good / service but on access to it, non-owners will enjoy greater flexibility by being able to invest in other products.
- From ownership to access. The Sharing Economy has made it possible to access the good / service without buying the property, in this way what was previously unattainable for some market segments, is now part of a new market accessible to all.
- Inflow of Venture Capital. The growth of the shared economy depends a lot on the subjects who invest in it: with the success of the Sharing Economy, many companies are committed to paying immense amounts in support of sharing start-ups, promoting their development.

The rise of the shared economy and its worldwide recognition is guided and permitted by the three factors described which are evolving. The sharing economy includes very different production and exchange models in terms of organizational forms and remuneration tools: some companies are oriented towards a logic of exchange and full reciprocity with the aim of generating a profit; others reconcile social aims and economic returns; still others look at sharing as an indirect reciprocity relationship, in which the counterpart is only possible.

Firms that adopt a logic outside the market and are oriented towards collaboration between peers and sharing, do not provide for remuneration or profits and sometimes not even the use of money (in some circumstances only the coverage of costs) nor the intermediation of profit-making entities. Other companies, on the other hand, operate in a market and profit logic, in which the platform earns from the brokerage activity, taking a commission on each transaction. In these cases, the "sharing" aspect derives from the adoption of innovative schemes, which place the user community at the centre, making use of the opportunities offered by new technologies and digital platforms, and above all from the adoption of optimization strategies use of shared resources. The operations, the nature of the platforms and the long-term impacts are configured by two variables: the orientation of the platform which can be profit or non-profit and the structure of the market: peer to peer or business to peer. The two variables cross in the table below; for each quadrant there is an example of a Sharing Economy platform.

Of course, each Sharing Economy company adopts a different business model, however, all have in common the fact that the initiative undertaken requires an organization, even minimal, which must be paid. Therefore, even initiatives that are completely free for users must generate a profit, to cover costs. Following a common classification, the most used models are the following:

- Service fee: the platform charges users by retaining a percentage of the monetary value of the exchanged service. The division between the two sides of the exchange of this tax varies for each platform: it can be entirely borne by those who use the good or consume the service (as for Gnammo), by those who produce it (such as Taskrabbit) or shared between the two subjects. (as for Airbnb).
- Subscription fee: the platform charges users for a fixed subscription, independent of the volume of transactions. The subscription model has two variants: freemium in which the basic service is free but the ancillary services are subject to a fee, tiered subscription in which the subscription has a different price for each level of use.
- Membership plus usage: a basic subscription payment is required to access the service and then you pay for the single use (as for car sharing or bike sharing services).
- White label: the platform offers a specific paid service for institutions, companies or organizations and, with the proceeds from this, keeps the service free for consumer users.
- Two-sided market: the platform offers the service free of charge to its customers thanks to the revenues it generates from the revenues from the advertising content on the site.

The Sharing Economy model turns out to be winning because it is based on great dynamism: in fact, final consumers are offered solutions that fully meet their needs. The sharing economy has profoundly changed the world market landscape, bringing some significant advantages: this can increase the efficiency of the economy by facilitating the meeting between supply and demand through the reduction of transaction costs and reducing the cost and the number of intermediaries thanks to increased competition; it can increase the total productivity of the factors allowing the use of capital goods by other users when they are not used by the owner; it can substitute the use and sharing of individual property; it can shift the emphasis from ownership towards access and encourage social interaction and community building; it can reduce the number of goods produced and reduce the volume of waste; it can reduce the environmental impact and waste.

Of course, the company that adopts a Sharing Economy model must include all relevant changes in the design of the model. In " key resources ", for example, the IT platform will be introduced, as it will allow transactions and take on the central role for the meeting of supply and demand. The " revenue streams " will depend heavily on whether and how the intermediary expects to generate profit from the user, the supplier or indirectly from third parties. Therefore, even the development of the Business Model Canvas will have to incorporate and contain the elements of the new economy in its design.

THE ADVENT OF THE CIRCULAR ECONOMY AS A SUSTAINABLE BUSINESS MODEL

The circular economy was born as a response to the crisis of the linear consumption model.

The linear economy is based on the extraction of raw materials, on mass production and consumption and on the disposal of waste once the end of the product's life has been reached. This incessant flow of inefficient and costly extraction and disposal has been, and still is, one of the main causes of phenomena such as marine and terrestrial pollution, the emission of greenhouse gases and the consequent climate change, as well as being the cause of bloody wars for the control of raw materials. Therefore, in a planet where available resources are limited, the linear consumption model, in addition to impacting the prices of inputs and consumer goods, does not even work from a geopolitical point of view. In the preface of "Towards the Circular Economy", Unilever CEO Paul Polman writes: "it is evident that an economy that extracts resources at increasing rates without considering the environment in which it operates, without considering the limits of our planet, cannot continue indefinitely ". Hence the need to implement an alternative development model: the enhancement of consumption waste, the extension of the life cycle of products, an economy of sharing resources, the use of recycled raw materials, the use of energy from renewable sources, can trigger a responsible way of producing and consuming, capable of improving the environmental conditions of our planet and those of the life of its inhabitants. The result of this research led to the development of a new, more sustainable model called the "circular economy".

The concept of circular economy is not easily traceable to an author or to a specific date, but it is the result of the union of different concepts and schools of thought. Already in the eighteenth century, thanks to Carlowitz, director of the royal office in the Kingdom of Saxony, who coined the term " sustainability ", the roots of the circular economy can be found. The theme of sustainability was then taken up again in 1798 by Malthus in his "Essay on the Population Principle", where the English economist observed that agricultural production, due to a fixed availability of resources, was subject to the law of diminishing returns and that therefore soon it would not be able to meet the geometric growth of the population. In 1982, the concept of circular economy was defined more precisely by Stahel with "The Product-Life Factor", where it is specified that the extension of the life span of a product, or the period in which it is used, it is a starting point for a smooth transition to a sustainable society and that extending the life span optimizes the resources used for production of a good, reduces the necessary natural resources and the resulting waste, thus generating growing well-being. A further step was added in 1997 thanks to Janine Benyus who coined the term "biomimicry", a new discipline that "studies the best ideas of nature imitating processes and mechanisms to solve human problems ". The biomimicry yes based on the idea that nature has already solved most of the problems we are experiencing facing. From here derive the three fundamental principles of the discipline (Benyus, 1997):

- Nature as Model: studying the models of nature to solve human problems;
- Nature as Measure: use ecological standards to assess the sustainability of innovations;
- Nature as Mentor: observe and evaluate nature based on what we can learn from it.

In 2002 William McDonough and Micheal Braungart published "Cradle to Cradle. Remaking the Way We Make Things ", in which the Cradle to Cradle emerges, that is, an approach to system design that consists in adapting industry models to nature, or converting production processes by assimilating the materials used to natural elements, which must therefore regenerate. It is therefore a question of adopting a perspective in which the social dimension and the environmental dimension coexist in an economic framework that intends to create systems that are not only efficient, but essentially environmentally compatible.

Finally, in 2010 the entrepreneur and economist Gunter Pauli, founder of ZERI Foundation (Zero Emissions Research and Initiatives) draws up the "Blue economy" report which theorises an economy model dedicated to the creation of a sustainable ecosystem by transforming previously wasted substances into profitable goods. It represents a development of the green economy: while the latter provides for a reduction of CO2 within an acceptable limit, the blue economy expects to reach zero CO2 emissions. Furthermore, the aim of the blue economy is not to invest more in environmental protection but, thanks to innovations in all sectors of the economy that use substances already present in nature, to make fewer investments, create more jobs. and achieve greater revenue.

The definition, therefore, is elaborated by the Ellen MacArthur Foundation, a non-profit foundation whose purpose is to accelerate the transition to a circular economy. The definition provided by the foundation outlines the circular economy as "a conceptually regenerative industrial economy that aims to allow effective flows of materials, energy, work and information so that natural and social capital can be rebuilt.

The innovation of business models turns out to be a fundamental step for companies operating in a sustainable scenario in which new economies such as Sharing and Crowd are born and evolve: it is necessary to found business models capable of capturing the new characteristics of such economies by evaluating and analyzing the new way in which the creation of value takes place, the management of resources and production capacities and the ways of accessing resources towards sustainability.

The birth of new and sustainable economies has brought with it the need to redefine the elements of the organization. The result of novelty was in fact radical: new actors and existing actors with new roles, new types of interactions, new types of products and services. At this point, it is necessary to define an innovative business model.

A business model can be shaped by analyzing the nine basic building blocks that encompass the four main areas of a business: customers, offering, infrastructure and financial strength.

The Business Model Canvas was proposed by Alexander Osterwalder in his first work, Business Model Ontology (2004), and then developed by Osterwalder himself, Yves Pigneur and Alan Smith together with a community of 470 experts in 45 countries around the world. a strategic Business Design tool that uses visual language to create and develop innovative business models, thus allowing to visually represent the way in which a company creates and distributes value for its customers. Today the Business Model Canvas is recognized as an international standard.

The elements arranged on the left of the model determine the efficiency of the organization; those placed on the right are the elements responsible for the creation of value.

At this point, we can note a certain similarity with the so-called "Blue Ocean" strategy as, with this, the company "swims" in a blue ocean, leaving the red one of ruthless competition. Even the "Blue Ocean" strategy is based on the logic of reducing costs for the company and increasing value for the customer in order to completely neutralize the competition. The value proposition, which occupies the central position in the Business Model Canvas, represents precisely the synthesis between these two objectives.

With regards to the sustainability, recent studies have developed circular business models (CBMs) to play attention to the topic above mentioned (e.g. reduce waste and turn waste into resources). The concept of circular economy has started to develop in response to the crisis of the traditional model, due to the need to deal with the limitations of resources used. The circular economy is an economy designed to "self-regenerate": materials of biological origin are intended to re-enter the biosphere, and materials of technical origin are designed to circulate within a flow with the least loss of quality. It is also an economy that intentionally yes "reconstitutes": aims to be based on renewable energy sources, to minimize, trace and eliminate the use of toxic chemicals, and to eliminate the production of waste and waste, through careful design.

An economy that aims to reduce the use of energy per unit of output and to accelerate the shift towards the use of renewable energy through design, treating everything that is present in the economy as a source of value. The idea goes beyond the requirements of the production and consumption of goods and services. The concept of the circular economy is based on the study of real, non-linear and feedback-rich systems, in particular of vital systems. " (Ellen MacArthur Foundation, 2013).

The circular economy model is based on the three " Rs ": Reduce, Reuse, Recycle. As we have already analyzed from the definition of Ellen MacArthur Foundation, it is a model that is based on the maximum exploitation of natural resources and is aimed at maximizing profit by reducing production costs. A circular approach aims at reviewing all stages of production. This focus passes through 3 fundamental principles indicated by the Ellen MacArthur foundation:

1. Preserve and increase natural capital by controlling limited deposits and balancing the flow of renewable resources;
2. Optimize the yield of resources through the circulation of products, components and materials of the highest value, at all times, in both cycles, biological and technical;
3. Identification and elimination of negative externalities that can discourage the effectiveness of the system.

According to the first principle, the system, when it needs it, must decide which resources to use and prefer the best processes and technologies that use resources renewables by balancing the flow.

The second principle is based on the distinction between biological and technical cycles. Biological cycles concern the management of all biological nutrients, i.e. renewable nutrients that can then be reintegrated into the biosphere so that with decomposition they return to being raw material for other subsequent cycles. Technical cycles, on the other hand, manage non-renewable materials that cannot be reintroduced into the biosphere, and which must therefore be designed to circulate for as long as possible. In both cases, but especially for technical cycles, restricted cycles are preferable, as they allow to conserve greater value and energy. This means designing not only to recycle, but also to regenerate and renovate.

Finally, the third principle emphasizes the importance of reducing or avoiding negative externalities such as air and water pollution, noise pollution and the release of toxic substances. These externalities create serious damage to the environment and discourage the effectiveness of the economic system.

The MacArthur Foundation based on the 3 fundamental principles of the circular economy it also lists some fundamentals that complete the overall picture of the model economic:

- The disappearance of waste: to maximize the circulation of products, components and materials, good design is essential, so that the product can be recovered and easily updated, minimizing the required energy input and maximizing the conservation of value.
- Building resilience through diversity: in a rapidly evolving context in which there is a high degree of uncertainty, resilience, understood as the ability of a system to deal with a shock, is built through flexibility and diversity. Specifically, a circular economy system to be resilient requires modularity, versatility and the ability to work with many different inputs.
- Systemic thinking: consists in understanding how the parts of a whole manage to influence each other. This ability is fundamental in the circular economy, because it allows you to analyze and design non-linear, evolutionary and feedback-rich systems. In this context, a further indication is inserted: to give priority to the effectiveness of the entire system and not to its efficiency. It is not necessary to maximize the efficiency of the single part if this had a negative impact on the whole system.
- Cascade actions: consists of using waste as an input to generate something new, "cascading" the value from discarded products to new products. To do this, it is necessary to identify the intersections between cycles and sectors in order to make the most of the potential provided by renewable matter.

The construction of a circular economy therefore has great strategic value both from an environmental point of view and from that of economic competitiveness. According to the European Commission, building a circular economy can save between 10% and 17% of primary resources every year, a percentage that can grow up to 24% by 2030 with the introduction of new production technologies and recycling. According to a study by the Ellen Mc Arthur Foundation, this could result in annual net savings for the manufacturing system of up to 640 billion dollars on the cost of procuring materials, equal to about 20% of the cost currently incurred. From an environmental point of view, the achievement of the recycling objectives indicated by the European Union would allow a further reduction of greenhouse gas emissions, in addition to the objectives already set, between 424 and 617 million tons. The companies that adopt this model will improve the efficient use of resources and redefine production processes, the nature of goods and services, waste management, consumption patterns and lifestyles, but also contribute to economic competitiveness and social inclusion, promoting the employment, equity and well-being of individuals.

Thus, implementing a circular business model (CBM) means rethinking the value creation, delivery and capture logic to implement circular strategies and bring economic benefits to the management field. Bocken (2015) adapts the business model canvas (next figure) to illustrate the peculiarity of a sustainable business model, and successively, Bocken, Schuit, and Kraaijenhagen (2018) adapt the discussion to the circular economy.

Figure 2. The circular business model (Adapted from Bocken, Schuit, and Kraaijenhagen, 2018)

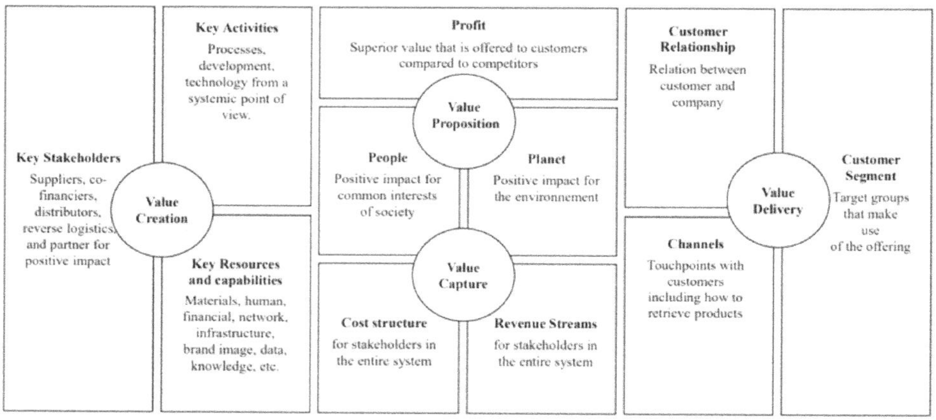

In explaining the creation, proposition, capture and delivery, Bocken, Schuit, and Kraaijenhagen (2018) identify six key elements of CBMs referring to the Canvas schema: a) reverse logistics is the main key activity of the entrepreneurial model, b) key resources are mainly material assets; c) one of the key activities concerns the process development ; d) people and planet are embedded in the formulation of the value proposition; e) costs and revenues are referred to the entire system and not the single firm; f) retrieving practices cover the value delivery.

Apart from changes in design elements of the business model canvas that offer a guide for practitioners who intend to frame the business model of circular projects, Bocken, Schuit, and Kraaijenhagen (2018) linger on the nature of CBMs. They define CBMs as uncertain, non-conventional, long range and multiple target models.

EXPLORATIVE ANALYSIS AND CASE STUDIES

Sharing Economy in Uber Health

Uber is the largest sharing economy company in the transportation sector, making its brand immediately recognizable and its business model a paradigm to follow. The platform works quite simply: users download their app, request a ride, share the car's license plate number that responds to the call, and the fare is automatically charged to the registered payment method upon arrival at the destination. Essentially, they function as private taxis bookable through smartphones, with taxi drivers as their primary competitors. Uber was founded in March 2009 in San Francisco and has since expanded globally, specializing in various sectors (Schneider, 2017). However, their easily replicable idea has led Uber to diversify its operations, including venturing into healthcare. On March 1, 2018, the new concept of Uber Health was born thanks to an article written by Chris Weber, CEO of Uber. Uber Health is a transportation service for doctors and patients, providing mobility for patients to access medical facilities easily and safely. For doctors, it means appointment coordination and the ability to schedule check-up visits in advance. Before the inception of Uber Health in 2018, the American Hospital Association reported that 3.6 million people missed medical appointments due to lacking transportation options. Uber Health, unlike other Uber services, can be accessed without an app by calling a number that sends a text message with the driver's details, offering an alternative for users unfamiliar with app interfaces, such as the elderly. Additionally, logistical challenges like accommodating wheelchairs and portable beds for severe patients have posed hurdles, given the specific vehicle requirements for Uber's services. Safety features are available 24/7 on the app and dashboard to assist users and drivers. The service offers flexible hours, giving drivers control over when and how much they work—Uber's unique selling proposition being "The more you drive, the more you earn." In Italy, Uber currently operates in only five cities (Rome, Milan, Florence, Turin, and Bologna), with Uber Health service not yet active. Government cuts to healthcare and public transportation could make Italy, especially the southern regions, a potential market for patients, doctors, and hospital staff who lack personal transportation options. While the market is becoming more automated, these are positive factors for Uber. However, Uber Health does not aim to replace emergency services like ambulances. Jay Holley, head of partnerships at Uber Health, emphasized that it is up to doctors and healthcare providers to evaluate each person's physical condition and determine if an Uber ride is the best mode of transport for them.

Being a work of horizontal integration, Uber's focus is therefore to fully exploit its existing platform which applies the same platform in three different sectors. Let us talk about Uber, Uber for Business and UberEATS. By explaining in more detail,

Uber aims to provide the patient with a more comfortable experience not only by guaranteeing them a transport, but also by granting a greater flexibility of schedules.

This is because Uber Business services help not only the patient, but above all the management of the clinic staff by allowing them, for example, to keep the clinic parking free or to organize shifts even at more extreme times such as early morning or late evening so to be able to give the opportunity to those who cannot free the selves from work to still be able to access medical examinations (Felländer, 2015).

To complete the work, therefore, Uber had to introduce a few changes that can be traced back to:

1) Compatibility with the Health Insurance Portability and Accountability Act of 1996 (HIPPA)14 which provides for the adoption of rigorous physical, network and procedural security measures for companies that process protected health data (PHI, Protected Health Information) inside the company's buildings.
2) "Technology should never be a barrier to healthcare." In the theoretical chapter it is highlighted how internet, access, and the ability to use modern technological means is not yet within everyone's reach, especially for example those who most of all need regular medical care and who at the same time could have economic difficulties or people more action. So, the whole service has been redesigned to be used without having a smartphone to use the application, but simply with an SMS or a phone call to the landline number.

The Case of Lyft: Creating Driving Program to Improve Patient Experience

Lyft is a private U.S.-based transportation company headquartered in San Francisco, California, founded in 2012. Like Uber, Lyft's core business involves providing rides to people in need of transportation, facilitated by private drivers who own their vehicles. Lyft also utilizes an app for ride bookings, positioning itself as one of Uber's main global competitors. Following the expansion of the largest sharing economy company, Lyft too has diversified its activities over the years, including entering the healthcare sector (Stephany, 2015). A major strength for Lyft is its partnerships with key players in the global healthcare ecosystem, relying on Lyft to provide reliable rides to their patients. Lyft has collaborations with 9 out of the top 10 U.S. healthcare systems and 9 out of the top 10 major non-emergency medical transportation (NEMT) brokers. Moreover, Lyft has also established a service called "Lyft Concierge," granting access to a network of over 2 million drivers. Their partners have observed that creating a driving program with Lyft can help reduce costs, increase efficiency, and enhance the patient experience. Through the Lyft Concierge dashboard, partners can immediately request rides for patients and schedule future rides. They can view ride details like pricing and driver routes, enhancing the overall patient experience. As stressed several times, transport is a problem especially for the weakest and most marginalized categories of society. From the pregnant patient with antenatal care appointments to the Medicare beneficiary, all complicated by the COVID-19 emergency. For this reason, Lyft introduced in April of this year the Pass for Healthcare whose objective is "put the power of Lyft into the patient's hand" thus putting the patient at centre of healthcare sector managing the business around his needs and not around the needs of doctors, health facilities or insurance companies (Chen et al., 2020). The goal compared to previous experiments is to give greater flexibility to the service to better adapt. The sponsoring healthcare or social services organization still covers the cost of the ride, while having access to built-in controls for budget, location, compliance, and more. If the analysis stopped at this point the sharing economy can only bring benefits to the sector, but we know that this business model has more than a few dark implications. It is necessary to check if it significantly solves the problem, if it has negative impacts on patients, if there is a risk for the free market and how this change has a negative impact on the working conditions of those who work in the NEMT sector.

The Role of Didi Chuxing in Revolutionizing Healthcare Transportation in China

Didi Chuxing, a leading Chinese ride-sharing company, has become a prominent player in the global transportation landscape. Founded in 2012, Didi rapidly expanded across China and globally, offering services ranging from traditional ride-hailing to carpooling, and even autonomous driving technologies. Didi's business model is based on leveraging its platform to connect drivers and passengers through a mobile app, with the company continuously enhancing its offering by introducing diverse services. One of the company's most notable innovations is its venture into the healthcare sector, specifically in providing transportation solutions for patients. This move exemplifies how ride-sharing platforms, particularly within the context of the sharing economy, can significantly impact industries beyond their traditional domain (Ma et al., 2019).

In 2017, Didi Chuxing launched its healthcare transportation service, an initiative aimed at addressing the growing transportation challenges faced by medical patients in China. The Chinese healthcare system, particularly in urban areas, faces significant logistical issues, including overcrowded hospitals, long waiting times, and limited access to transportation for patients, especially those living in rural areas. For many patients, especially the elderly, accessing healthcare services can be a monumental challenge due to lack of reliable transportation. These barriers have led to patients missing medical appointments, delayed treatments, and overall dissatisfaction with the healthcare system. Recognizing this issue, Didi introduced a service to bridge the gap between patients and medical facilities.

Didi's healthcare transportation service functions by allowing healthcare providers to arrange rides for their patients directly through the Didi platform. This service is particularly valuable in a country like China, where the rapid urbanization has created massive demand for healthcare services, yet many areas still lack adequate infrastructure for transportation. Didi's model offers flexibility for patients, as they can book rides tailored to their specific medical needs, including wheelchair accessibility and specialized vehicles. The company also provides an option for healthcare providers to schedule rides in advance, allowing for greater planning and reducing wait times for patients who require urgent medical attention (Shao et al. 2022).

Importantly, Didi's service targets not only patient transportation but also addresses the needs of medical staff and healthcare facilities. Through partnerships with hospitals and healthcare systems, Didi offers ride-sharing options for healthcare professionals to commute to and from work. This can help alleviate the burden on hospital parking facilities, particularly in densely populated urban centers, and also provides staff with a more reliable and efficient means of commuting. This aspect of the service underscores the value of Didi's transportation model in the healthcare sector, as it contributes to the efficient functioning of medical institutions by optimizing staff movement and minimizing delays.

One of the key advantages of Didi's healthcare-focused platform is the ease of access. Similar to other platforms in the sharing economy, Didi offers users an intuitive app to book rides. However, acknowledging that not all patients may be tech-savvy, especially older adults or those in rural regions, the service has been designed to accommodate a wide range of users. For patients who may not have smartphones or are unfamiliar with app interfaces, Didi has integrated call-based booking systems. This feature ensures that even the most vulnerable populations can access the service without being excluded due to technological barriers.

Despite these benefits, the integration of Didi into the healthcare transportation ecosystem has not been without challenges. One issue that has been raised is the potential for overcrowding in an already strained healthcare system. While Didi's platform offers convenient access to transportation, it does not address the root causes of transportation shortages, such as infrastructure gaps in rural areas or limited healthcare services. Moreover, like other companies in the sharing economy, Didi faces scrutiny regarding the working conditions of its drivers, who may experience varying levels of income stability, especially in sectors where demand can fluctuate significantly. In the healthcare space, ensuring that drivers are adequately compensated for providing timely, reliable, and often sensitive transportation services to patients is a crucial concern that Didi must address (Ko, 2019).

Furthermore, the widespread implementation of Didi's services raises broader questions about the long-term sustainability of such business models. While ride-sharing platforms like Didi offer flexibility and convenience, they also create potential risks to public healthcare funding. If private companies dominate the transportation of patients to medical facilities, there may be concerns over equity, as those with fewer resources may find it increasingly difficult to access essential health services. Additionally, as with other sharing economy models, the displacement of traditional workers (e.g., taxi drivers or public transport employees) could create significant social and economic challenges.

CONCLUSION

In conclusion, Didi Chuxing's healthcare transportation service represents a transformative step within the sharing economy, highlighting how platforms traditionally associated with transportation can be leveraged to address healthcare inefficiencies. By enabling easier access to medical facilities for patients and improving transportation options for healthcare professionals, Didi provides valuable support to China's healthcare sector. However, it also raises important considerations about equity, worker rights, and the broader impact of integrating private enterprises into essential public services. As the sharing economy continues to evolve, Didi's approach will likely serve as a model for future integrations between technology-driven companies and critical sectors such as healthcare, underscoring the need for careful management of value co-creation and long-term sustainability.

REFERENCES

Audretsch, D. B., Braunerhjelm, P., & Carlsson, B. (2005). *Growth and entrepreneurship: An empirial assessment* (No. 3205). Papers on Entrepreneurship, Growth and Public Policy.

Becattini, G. (1999). La fioritura della piccola impresa e il ritorno dei distretti industriali. *Economia e Politica Industriale*, *1999*(103).

Benyus, J. M. (1997). *Biomimicry: Innovation inspired by nature* (Vol. 688136915). Morrow.

Bocken, N. M., Schuit, C. S., & Kraaijenhagen, C. (2018). Experimenting with a circular business model: Lessons from eight cases. *Environmental Innovation and Societal Transitions*, *28*, 79–95. DOI: 10.1016/j.eist.2018.02.001

Boer, H., Berger, A., Chapman, R., & Gertsen, F. (2017). *CI changes from suggestion box to organisational learning: Continuous improvement in Europe and Australia*. Routledge.

Carayannis, E. G., & Campbell, D. F. (2009). 'Mode 3'and'Quadruple Helix': Toward a 21st century fractal innovation ecosystem. *International Journal of Technology Management*, *46*(3-4), 201–234. DOI: 10.1504/IJTM.2009.023374

Chen, K. L., Brozen, M., Rollman, J. E., Ward, T., Norris, K., Gregory, K. D., & Zimmerman, F. J. (2020). Transportation access to health care during the COVID-19 pandemic: Trends and implications for significant patient populations and health care needs.

Felländer, A., Ingram, C., & Teigland, R. (2015). Sharing economy. In *embracing change with caution. Näringspolitiskt Forum Rapport* (Vol. 11).

Grant, R. M. (1999). Transforming uncertainty into success: The strategic leadership forum 1999. *Strategy and Leadership*, *27*(4/5), 31–36. DOI: 10.1108/eb054645

Haeckel, S. H. (1999). *Adaptive enterprise: Creating and leading sense-and-respond organizations*. Harvard business press.

Iaione, C. (2016). The CO-City: Sharing, collaborating, cooperating, and commoning in the city. *American Journal of Economics and Sociology*, *75*(2), 415–455. DOI: 10.1111/ajes.12145

Ko, S. (2019). *Didi Chuxing: expansion and risk management*. SAGE Publications: SAGE Business Cases Originals.

Kramer, M. R., & Porter, M. (2011). *Creating shared value* (Vol. 17). FSG.

Lazonick, W. (2007). The US stock market and the governance of innovative enterprise. *Industrial and Corporate Change*, *16*(6), 983–1035. DOI: 10.1093/icc/dtm030

Leydesdorff, L. (2012). The triple helix, quadruple helix,..., and an N-tuple of helices: Explanatory models for analyzing the knowledge-based economy? *Journal of the Knowledge Economy*, *3*(1), 25–35. DOI: 10.1007/s13132-011-0049-4

Leydesdorff, L., & Etzkowitz, H. (1996). Emergence of a Triple Helix of university—industry—government relations. *Science & Public Policy*, *23*(5), 279–286.

Ma, L., Zhang, X., Ding, X., & Wang, G. (2019). Risk perception and intention to discontinue use of ride-hailing services in China: Taking the example of DiDi Chuxing. *Transportation Research Part F: Traffic Psychology and Behaviour*, *66*, 459–470. DOI: 10.1016/j.trf.2019.09.021

MacArthur, E. (2013). Towards the circular economy. *Journal of Industrial Ecology*, *2*(1), 23–44.

Marquesan, F. F. S., Zawislak, P. A., & Camboim, G. F. (2024). The Functions of the Proto-entrepreneur in Building up Startup Innovation Capabilities. *BAR - Brazilian Administration Review*, *21*(4), e230076. DOI: 10.1590/1807-7692bar2024230076

Mascarenhas, O. A., Kesavan, R., & Bernacchi, M. (2006). Lasting customer loyalty: A total customer experience approach. *Journal of Consumer Marketing*, *23*(7), 397–405. DOI: 10.1108/07363760610712939

Nosratabadi, S., Mosavi, A., Shamshirband, S., Zavadskas, E. K., Rakotonirainy, A., & Chau, K. W. (2019). Sustainable business models: A review. *Sustainability (Basel)*, *11*(6), 1663. DOI: 10.3390/su11061663

Rashid, A., Asif, F. M., Krajnik, P., & Nicolescu, C. M. (2013). Resource conservative manufacturing: An essential change in business and technology paradigm for sustainable manufacturing. *Journal of Cleaner Production*, *57*, 166–177. DOI: 10.1016/j.jclepro.2013.06.012

Roome, N., & Louche, C. (2016). Journeying toward business models for sustainability: A conceptual model found inside the black box of organisational transformation. *Organization & Environment*, *29*(1), 11–35. DOI: 10.1177/1086026615595084

Rullani, E. (1999). La diversità italiana alla prova del postfordismo. *Economia e Politica Industriale*, *1999*(103).

Schneider, H. (2017). *Creative destruction and the sharing economy: Uber as disruptive innovation*. Edward Elgar Publishing. DOI: 10.4337/9781786433435

Shao, H., Jin, C., Xu, J., Zhong, Y., & Xu, B. (2022). Supply-demand matching of medical services at a city level under the background of hierarchical diagnosis and treatment-based on Didi Chuxing Data in Haikou, China. *BMC Health Services Research*, *22*(1), 354. DOI: 10.1186/s12913-022-07762-4 PMID: 35300679

Stephany, A. (2015). *The business of sharing: Making it in the new sharing economy*. Springer. DOI: 10.1057/9781137376183

Vargo, S. L., & Lusch, R. F. (2004). Evolving to a new dominant logic for marketing. *Journal of Marketing*, *68*(1), 1–17. DOI: 10.1509/jmkg.68.1.1.24036

Chapter 4
Value Co-Creation in Neuromarketing Strategies:
An Explorative Analysis

ABSTRACT

The concept of neuromarketing is based on the application of recent innovations acquired in the neuroscience field in order to investigate consumer behavior, exploring not only conscious decision-making processes, but also unconscious ones. While traditional market research has always focused mainly on the rational sphere of decisions, neuroscience has shown that most purchasing decisions occur in an irrational and unconscious way. In practice, emotions, feelings, implicit associations and processes related to unconscious habits are responsible for most decisions. Thanks to technological tools such as functional magnetic resonance imaging (fMRI), electroencephalogram (EEG) and eye-tracking, neuromarketing is able to provide valuable data on brain processes that consumers themselves are not aware of. This allows companies to develop more effective marketing strategies, based on a deep understanding of unconscious emotional and cognitive reactions, which strongly influence purchasing behavior.

INTRODUCTION

Neuromarketing: Historical and Theoretical Aspects

Ale Smidts, a marketing research professor, in 2002, gave the name to this new discipline, neuromarketing: "the set of techniques for identifying brain mechanisms aimed at a greater understanding of consumer behavior for the development of more

DOI: 10.4018/979-8-3373-3466-0.ch004

effective marketing strategies." Neuromarketing is, according to most scholars and economists of our decade, the most innovative model to explain to the company how the consumer's mind works, to bring out the feelings, needs and conscious and unconscious preferences of man in the eyes of the company (Utriainen, 2020). This new science, through the fusion of economic, psychological and social studies, acts as a bridge between what the customer wants and what the company offers. This happens by analyzing the crucial moment of the commercial exchange for this branch of marketing, that is, the moment in which the consumer receives the company communication, in the form of advertising, packaging, sensorial stimuli and offers the company a response regarding the advertising stimulus. "It is in the moment of the trigger that much of the success or failure of a marketing strategy is summarized today. Stimulus = response: this is the new equation of neuromarketing [...] triggers offer marketing the possibility of quickly reaching the minds of customers and if they are effective, they can guarantee a notable competitive advantage, because we are no longer just talking about communication but about the possibility of creating new habits, routines and new consumer behaviors linked to the brand." (Francesco Gallucci, "Neuromarketing"). Brands that manage to find a link between internal and external stimuli find a new key to success, an example are social media such as Instagram, Tik Tok, Facebook that have filled a gap in consumer needs, have created a new habit in him and have been able to unify a feeling, boredom, with the need to distract oneself with these applications.

On the other hand, environmental, social and economic growth models can be achieved in neuromarketing strategies through the implementation of the 2030 Agenda for Sustainable Development, an action program consisting of 17 Goals outlined by the UN General Assembly on 25 September 2015.

For instance, recycling is a concept similar to the idea of zero waste and for this reason it is considered valid, because it makes the waste from a production cycle, raw material for a new cycle, a fundamental element for the circular economy and represents one of the fundamental tools for transforming value chains from an eco-sustainable perspective, contributing to economic and environmental sustainability. Neuromarketing approaches help firms rethink this process by analysing circular business models through the study of messages and responses among different actors involved. Furthermore, interventions to support the circular economy, a key factor in green logistics, have taken on growing importance within the National Recovery and Resilience Plan (PNRR), which foresees factors such as quality, punctuality in planning, and sustainability.

For example, using PaperLab, the Epson system dedicated to the dry recycling of paper in the office, companies can recycle used paper directly in the office, without polluting transport. Reusing used paper means reducing the environmental impact caused by greenhouse gas emissions, since its production process is decidedly more efficient than that of wastepaper production.

Neuromarketing studies communication and therefore the triggers between a message and a response, so that the company can learn essential information for its own market strategies such as:

1) Who views the product.
2) The product remains in the consumer's mind.
3) Which elements of the product attract the consumer the most.
4) How long does a customer watch an advertisement?

To answer these questions of the company, this science uses hyper-technological and innovative tools, through which it can literally enter the consumer's brain and understand, through glasses, specialized helmets, applications and data exchange, to analyze what an individual's eyes focus on most when observing a billboard or what pushes an individual to overtake or, on the contrary, to linger on an advertisement. Neuromarketing tracks eye movement, measures brain activity and human emotions, transforms this information into market analysis aimed at improving the company performance of a product, sales and the impact that the company produces in the mind of the consumer (Glaenzer, 2016). Marketing through this study model attempts to explore the customer's evaluation and decision-making process on a product before purchasing, sometimes trying to "trick", not fraudulently, the individual with communication techniques that derive from the principles of neuromarketing. In the world, today, hundreds of relevant companies on the market invest their research and development budgets in this new science, among these we can mention Coca-Cola, Mc Donald, Apple, Google, Ferrero, Barilla etc.

In order to better understand the above-mentioned technique, it is effective to report some practical demonstrations of neuromarketing application in companies:

An explanatory example of neuromarketing, which allows us to grasp the real effectiveness and operation of this new science, is the Microsoft company, which analyzed how the advertising campaign, related to its "Xbox" platform, stimulates and interacts with the consumer. The company carried out evaluation tests that monitored human parameters such as heart rate and breathing, brain activity, gaze and head movements, the time between closing and reopening the eyelids, skin temperature, etc., all this while watching three 30-second interactive commercials on the platform. The study, through eye tracking, was carried out to understand which of the three advertisements was more engaging and impactful and showed that there were three different modes of involvement and attention, some more amplified thanks to memories and emotions for the brand specific to the individuals in question. "Through the study, it was therefore possible not only to measure the effectiveness of advertising across all types of media but also to explore how neuroscientific technologies can help answer two questions that marketers have been asking for years: how to measure the audience's engagement with their brand and how to measure the impact of advertising across different types of media," says Microsoft (Rawnaque, 2020).

- At the end of the last century, the soft drink company PepsiCo devised a strategy to collect data on customer perceptions and preferences through an experiment: it had a wide range of people taste both a Pepsi and a Coca-Cola brand drink, with the company logos masked. Consumers, for the most part, surprisingly preferred Pepsi, given that Coca-Cola has always boasted a larger market share. Teach Carilion, a neuroscientist from Virginia at the Research Institute, wanted to understand why Pepsi was not then a leader in the carbonated soft drink sector and connected 67 people to a CT scanner and an MRI, giving them a taste of both drinks. The experiment showed that, when they drank, the areas related to the brain's reward system were activated and above all that when they were shown the brands of the drinks, the medial prefrontal cortex, the area of the brain that deals with higher thought, was activated, making it clear that it generated a series of images, sensations and feelings in their minds. These emotions are what push the final consumer to prefer a perfectly replaceable product to another, and the memory that was born, linked to the image of the Coca-Cola brand, in the minds of those individuals, was much more impactful and significant.

For instance, neuroeconomics and neuromarketing are interdisciplinary disciplines that combine psychology, neuroscience, economics and marketing to better understand how humans make economic decisions and how they react to marketing messages. These scientific fields offer an innovative perspective, trying to explain decision-making processes and consumer preferences through the analysis of the brain and the physiological responses of the organism.

Neuroeconomics was officially born in the early 2000s, thanks to the intuitions of two Israeli psychologists, Daniel Kahneman and Amos Tversky – generally considered the founders of the so-called behavioral economics – and the concomitant discovery of innovative technological tools for the visual representation of brain functions.

The theories of these two scholars, based on the idea that "the human mind tends to fall into error when forced to choose in situations of uncertainty", in addition to having laid the foundations of this new discipline, were awarded the Nobel Prize won by Kahneman in 2002.

The assumption from which neuroeconomics starts is that - unlike what traditional economics claims - man is not a rational animal but rather acts under the impulse of automatic and very often unconscious neuronal processes, therefore independent of the will of the individual. This means that economic behavior is the result of a sort of negotiation between rational and emotional elements, or between instinctive and automatic aspects and conscious and rational aspects of personality.

The main objective of neuroeconomics is therefore to integrate neuroscientific models with economic models to explain the mechanisms underlying economic behavior, such as choice and risk.

From a theoretical point of view, while paradigms deriving from classical economics, such as the homo economicus model, postulated that individuals act in a perfectly rational way to maximize their benefits, neuroeconomics, on the other hand, proposes that decisions are influenced by automatic processes, emotions and unconscious impulses, questioning the foundations of rational theories. To confirm its intuitions, neuroeconomics uses models based on emotions, implicit preferences and the influence of dopamine in the brain during the evaluation of rewards. Among the tools used to corroborate its research, neuroeconomic scholars use some technologies such as fMRI, electroencephalogram (EEG) and eye-tracking techniques to study reactions to economic stimuli.

Neuromarketing also began to spread in the early 2000s, as a practical application of neuroeconomics in the field of marketing. The discipline has attracted great interest thanks to its ability to analyze consumers' unconscious responses to advertising stimuli (Nyoni and Bonga, 2017).

From a historical point of view, according to specialized literature, the term neuromarketing was first coined on October 25, 2002, by Dutch professor Ale Smidts during a speech given on the occasion of the awarding of the chair of Marketing Research Methodology at the Erasmus University of Rotterdam.

The birth of the term neuromarketing, as Ale Smidts himself recently recalled during an international conference[1], is linked to his intention to exploit the possibilities offered by electroencephalography (EEG) to study the brain's response to an advertising spot, with an analysis of the scenes most remembered by participants. Starting from this idea, the Dutch researcher began to study the scientific literature on the topic (i.e. that relating to neuroscience) to try to identify other methods and tools useful for the study of the consumer and for the study of advertising stimuli, taking into consideration in particular positron emission tomography (PET) and functional magnetic resonance imaging (fMRI).

In 2004, the American neuroscientist Read Montague[2] subsequently proposed what is instead considered the first empirical neuromarketing research. It is an original experiment to study the preferences between Coca Cola and Pepsi Cola in consumers (Vecchiato, 2013).

In extreme synthesis, the two different drinks were offered in anonymous glasses. Before entering a machine for the *Magnetic Resonance Imaging*[3], they were made to taste the two drinks. Afterwards, the participants were made to enter the machine and were asked to think about the moment – a few moments before – in which they had tasted the drinks from the two glasses.

The goal, as the scholar himself explained, was to demonstrate how people do not always make rational choices and, above all, how emotions and memories often influence our choices in a decisive way. The experiment, in practice, had among its main objectives precisely that of investigating a little more deeply what happened at a neural level in the brains of people who found themselves having to make this simple choice: Coca Cola or Pepsi Cola?

The result was that over 50% of people preferred Pepsi, confirming the results of a similar study, already carried out in 1975, when however, the technologies for analyzing brain function invented in the meantime were not yet available. This time, in short, thanks to tools such as Magnetic Resonance, there was the possibility of obtaining valuable additional information compared to thirty years earlier.

The data showed that, in fact, at the time of anonymous tasting (i.e. when people did not know where Pepsi was and where Coca Cola was), the area of the brain that lights up when we like a flavor was activated.

However, if, before tasting, the interviewees were also told the brand of the drink they were about to drink (so they were told what they were drinking, at the moment they were drinking it), in their brains, in addition to the taste area as before, the judgment area was also activated (based on the rational component of the brain and not on the emotional one).

In practice, at a cerebral level a conflict was generated between the area of emotional pleasure and that of rational judgment and in many cases Coca Cola prevailed (Morin, 2011).

This was proof that Coca Cola's supremacy in the soft drink market was based on a purely emotional component, which meant that the image and values of the brand prevailed over the individual benefit generated by the consumption of the product itself. (Benefit, in this case, to be understood as the neuronal well-being linked to pleasure and taste. That is, I drink what is most pleasant for our palate).

As can also be seen in the example just mentioned, the main objectives of neuromarketing, in extreme synthesis, are to understand how emotions and unconscious processes end up influencing consumer preferences and improve marketing strategies based on this information.

Before resuming and deepening the analysis of the most recent methods that have been developed for this purpose in the field of neuromarketing research, it is particularly important to consider in more detail its theoretical foundations which are rooted, as mentioned, in psychology and behavioral economic theories developed during the XX century.

Behaviorism and cognitivism are two of the main schools of thought in psychology, which emerged in distinct periods and with different approaches to the study of human behavior. The psychological theories produced by these approaches have significantly shaped the history of marketing. While behaviorism laid the foundations for behavioral reinforcement techniques, cognitivism later allowed for an in-depth analysis of decision-making processes.

HUMAN BEHAVIOR AND COGNITIVE NEUROSCIENCE: MAIN THEORIES AND VALUE CO-CREATION ASPECTS

Behaviorism was born in the early decades of the 20th century. Its beginning is in fact set in 1913, the year in which John B. Watson published the famous article Psychology as the Behaviorist Views It, considered the manifesto of behaviorism. The theory would later develop during the 1920s and 1930s and would consolidate with the contribution of B.F. Skinner in the 1940s and 1950s (Foxall, 1987).

Behaviorist psychologists focused mainly on observable and measurable behavior, almost completely excluding the role of mental processes within the individual, since they were considered inaccessible to scientific research. According to behaviorists, human behavior can be simply understood and explained as a series of responses to environmental stimuli.

One of the main representatives of this school is certainly the Russian Ivan Pavlov (1901-1920), who founded the so-called theory of Classical Conditioning.

Pavlov is famous for an experiment conducted with some of his dogs. The scientist realized that these, especially when they were hungrier, began to salivate even before he presented them with food. Pavlov then inserted a tube into the oral cavity of one of his dogs, which began to signal salivation when a piece of meat was shown to it (Adams, 2020).

At this point the scientist wanted to experiment more deeply with psychic secretion: he rang a bell a few times, which did not provoke any reaction in the dog, except for a pricking up of the ears. Then Pavlov rang the bell and subsequently gave the animal the slice of meat; therefore, the dog associated the sound with the imminent arrival of the steak.

Finally, by repeating the stimulus several times, the scientist observed that the level of salivation increased as soon as the bell rang, well before the meat was presented to it.

The sound of the bell had become, in Pavlov's terminology, a conditioned stimulus and the consequent increase in salivation at the sound of the bell was a conditioned response.

The theory that followed was that the animals (it was later understood that the experiment could be generalized to humans) learn to associate one stimulus with another, regardless of the specific characteristics of the stimuli. In "classical" (or Pavlovian) conditioning, reflexes, that is, unknown and uncontrollable responses, are the reaction to stimuli: such as watering one's mouth when one sees or smells food, contracting the pupils when they are exposed to a light source, closing the eyes when an object approaches quickly: it is something that each of us does completely unconsciously.

Pavlov's research, in short, allowed us to understand the dynamics of associative learning through the phenomenon of classical conditioning.

Four distinct elements that characterize classical conditioning were subsequently defined:

Unconditioned stimulus: It causes a significant response for the subject and is represented by the food that manages to make the dog salivate.
Conditioned stimulus: Also called neutral, because initially it does not generate any response in the subject, but through the association with the stimulus, the bell generates a response: salivation.
Unconditioned response: It is the subject's response to the unconditioned stimulus. In the experiment it is represented by the production of saliva caused by the sight of food.
Conditioned response: It is the subject's response to the conditioned stimulus, which in this case is the dog's salivation at the sound of the bell.

The fundamental principle, to summarize, is that of the unconscious association of a neutral stimulus with an involuntary response.

Later, other scholars explored this principle in greater depth, such as with the so-called Operant Conditioning of Skinner, in the mid-twentieth century and the Theory of Social Behavior of Albert Bandura, in the 1960s.

Skinner demonstrated that a rewarded behavior (positive reinforcement) is more likely to be repeated, while a behavior followed by punishment (negative reinforcement or punishment) tends to decrease; Bandura expanded behaviorism to include observational learning or modeling, demonstrating that people learn by observing the actions of others and their consequences.

All these versions of the behaviorist approach will have multiple possibilities of application in the field of marketing. In fact, purchasing decisions do not depend only on the functionality of the product, which sometimes take a back seat to the sensations and emotions that are felt at a given moment. By intercepting the decision-making process of a consumer, it will be possible to lead him to complete the purchase. Or it is possible to raise so-called latent or unexpressed needs of the consumer who will find himself buying following the impulse.

Just to mention a few examples, advertising uses classical conditioning to associate positive emotions with products, such as when a brand is associated with pleasant images to create a positive emotional response in consumers; loyalty techniques, such as points and rewards programs, are based on operant conditioning: consumers are "rewarded" for their purchases, encouraging the repetition of the purchasing behavior. Finally, the theory of social behavior is also fundamental in those advertisements in which consumers see behavioral models (testimonials, influencers) who act as examples to emulate.

An Example: Olfactory Marketing

The *Dunkin Donuts* campaign is a very famous olfactory marketing campaign[4]: to improve the sales of their coffee, in several buses in Korea, some nebulizers with an intense coffee aroma were installed while their jingle was played on the radio, so as to involve the senses of people who were going to work or school (Harizi, 2021).

To top it all off, when passengers got off the bus, they found themselves in front of an advertising poster with the Dunkin Donuts brand. In this way, the brain reacts to the different stimuli proposed with a positive result for coffee sales, which improved by 29% and visits increased by almost 20%.

Not only aroma diffusers in buses, but also billboards are perfect for these campaigns: *Cacaolat*, a company born in Barcelona and has become famous for its delicious chocolate milkshake, has had enormous success thanks to a sensorial marketing campaign.

How? He invited people to experience the unmistakable chocolate aroma of his smoothies, attracting them to the billboards from which the intense and delicious smoothie scent was released. A unique experience that was very successful among passers-by.

Cognitive Psychology

Cognitive psychology emerged as a response to the limitations of behaviorism in the 1950s and 1960s, with the beginning of the so-called "Cognitive Revolution" in psychology. The publication of Cognitive Psychology by Ulric Neisser in 1967[5] marked the consolidation of cognitivism as the mainstream in the study of human behavior in psychology.

Cognitive psychology considers internal mental processes essential to understanding human behavior, including elements such as perception, memory, thought, and language.

Cognitive processes are seen as the way in which people interpret and use information, which is different from behaviorism, which ignores these aspects (Kleinginna Jr and Kleinginna, 1988).

In short, while behaviorists tended to reduce human behavior to stimuli and responses, neglecting internal processing processes (through the study, for example, of perception, memory, linguistic processing and decision-making processes) and the sometimes much more complex and indirect reasons that lie behind choices, cognitivism aims to analyze the "why" of certain behaviors that are only apparently inexplicable, introducing the idea that individuals do not respond passively to stimuli, but actively process the information received through external stimuli.

One of the first cognitive theories is known as Information Processing Theory and has been developed since the 1950s. In this model, the human mind is compared to a computer in which information is collected, stored, processed and used to make decisions. The study of memory and decision-making processes is therefore central to this theory.

Another important theory, developed by Albert Bandura himself, whom we have already encountered among the behaviorists, is the so-called Observational Learning Theory. It is based on the concept of observational learning and maintains that individuals learn by observing the actions and consequences of the actions of others and can be interpreted from both a behaviorist and a cognitive point of view.

Leon Festinger's Cognitive Dissonance Theory maintains that when a person has conflicting thoughts or beliefs, psychological discomfort emerges, and people are driven to reduce this dissonance by changing their beliefs or behaviors.

A more recent theoretical model of a cognitive nature is Prospect Theory (Kahneman and Tversky, 2013) which tends to explain how people make decisions under risk conditions, often using an irrational approach. It shows that people evaluate losses and gains asymmetrically (losses have a greater impact than gains).

Understanding cognitive processes has generally allowed the development of more complex marketing strategies than behaviorism, based on the influence of consumers' thoughts, emotions and decisions. For example, marketing today uses techniques to influence perception and attention and uses so-called cognitive biases to guide decisions.

The information processing theory is used to understand how consumers process information about products (for example, companies try to simplify advertising messages to facilitate memory and understanding); the Observational Learning Theory emphasizes the importance of role models in advertising messages (such as celebrities or ordinary people) and their influence on the choices of consumers who observe and tend to emulate such models; the Cognitive Dissonance Theory is used when a product is presented as a "better" choice that resolves an internal conflict of the consumer (for example: "make a healthy choice"); finally, the Prospect

Theory is used in marketing to create offers that minimize the perception of risk. An example is the use of "money-back guarantees" or the framing of a product as a "can't-miss opportunity."

In conclusion, the shift from behaviorist to cognitivist theories represented a fundamental change in the approach to the study of human behavior and, consequently, to marketing research. Both certainly had a significant impact on a deeper understanding of human behavior: behaviorism provided tools to condition behaviors, while cognitivism opened the door to understanding the mind of the consumer. Today, marketing scholars tend to integrate aspects of both theories, and with the advent of neuromarketing, it is able to further focus into how people process and respond to advertising messages.

NEUROMARKETING: METHODOLOGICAL ASPECTS AND THEORETICAL SPECULATIONS

Neuromarketing – as we have already mentioned – is a field of research that tends to integrate the most recent neuroscientific discoveries with economic and psychological theories useful in order to scientifically understand and explain the unconscious processes and brain reactions of consumers to marketing stimuli (Spence, 2019). Therefore, it is based, as we have seen, on various scientific and psychological theories, which come from both neuroscience and cognitive psychology, and uses methods such as fMRI, EEG, eye-tracking and biometry to directly observe the brain and nervous system during exposure to advertising stimuli.

The main scientific and psychological theories that support neuromarketing and some of its key theoretical applications are the following:

Theory of Resonance and Emotional Contagion

The human brain is predisposed to react to the emotions of others through the so-called mirror neurons, discovered by the team of neuroscientist Giacomo Rizzolatti during the 1990s. Mirror neurons activate the same areas in our brain that are involved when we observe someone expressing an emotion or performing an action.

This theory is used in marketing to create advertisements that evoke observable emotions (such as happiness, surprise or empathy), making the viewer "infected" by the emotions of the advertising character feel closer to the brand.

Reward and Dopamine System Theory

Neuroscience has shown that the brain's reward system, particularly linked to the neurotransmitter dopamine, is active when we experience pleasure and gratification. This system is activated not only by physical rewards, but also by anticipation of reward (such as the anticipation of a purchase).

Neuromarketing techniques seek to stimulate the reward system to increase motivation to purchase. For example, advertisements that leverage the desire and anticipation of owning a product can activate these brain areas, making the purchase more likely.

Decision-Making Theory and Bounded Rationality

According to the theory of bounded rationality, developed by Herbert Simon and later expanded upon by Daniel Kahneman and Amos Tversky with Prospect Theory, humans do not make purely rational decisions. People are influenced by cognitive biases and emotional factors, especially when faced with uncertainty.

Neuromarketing experts use this theory to shape advertising campaigns, considering consumers' cognitive biases, such as the framing effect, where perception changes depend on how an option is presented. For example, a promotion that emphasizes the "loss" of not purchasing a product may be more effective than one that emphasizes the gain.

Maslow's Hierarchy of Needs and Motivation Theory

The Hierarchy of Needs Theory, formulated by Abraham Maslow, states that human needs are organized in a hierarchy, from basic physiological needs to self-actualization. People tend to satisfy lower needs before focusing on higher ones.

This theory guides companies in segmentation and targeting, helping them position their products as solutions to specific needs. For example, a luxury brand may appeal to needs for esteem and belonging, while a food product may focus on basic physiological needs.

Selective Attention Theory

This theory is based on the idea that the human mind can only process a limited number of stimuli at a time; therefore, it highlights the importance of capturing and maintaining the consumer's attention in an environment of information overload.

Eye-tracking techniques are used to understand which visual elements capture the viewers' attention the most. This data helps optimize the visual layout of advertisements, highlighting the points of greatest interest.

Theory of Emotions in Decision Making (Somatic Marker Hypothesis)

The famous theory of Portuguese neuroscientist Antonio Damasio argues that emotions play a key role in decisions. Previous emotional experiences create "somatic markers" that influence future choices.

Neuromarketing experts use emotion monitoring to design brand experiences that create lasting positive emotions, influencing purchasing decisions. For example, emotional campaigns can create an emotional bond between consumers and brands, increasing loyalty.

Priming Effect and Cognitive Biases

Priming is a psychological phenomenon whereby an individual is unconsciously influenced by previous stimuli, which condition the perception and reaction to subsequent stimuli. Cognitive biases, on the other hand, are distortions of reasoning that influence decisions.

Priming is used in neuromarketing to influence consumer preferences in a slight way. For example, a brand can use colors and symbols associated with positive values to predispose consumers to positively welcome the advertising message.

Brand Love and Emotional Loyalty Theory

This theory explores the emotional bond between the consumer and the brand, arguing that brand love can generate strong and lasting loyalty. It is a relationship that goes beyond rationality and is based on emotional experiences.

Companies use campaigns aimed at eliciting intense and positive emotions related to the brand, thus strengthening loyalty. Experiential events, emotional storytelling and testimonials are some of the tools to build brand love.

Visceral and Biometric Response Model

So-called visceral responses, such as heart rate and skin conductance, are automatic reactions that can indicate the emotional intensity of a consumer's response to a marketing stimulus.

Through biometrics (physiological measurements), neuromarketing evaluates which advertising elements stimulate stronger emotional responses. This helps marketers identify which aspects of a campaign are most effective in generating emotional engagement.

Optimal Experience and Flow Theory

Derived from Mihaly Csikszentmihalyi's positive psychology, this theory argues that people reach a state of flow when they are fully immersed and engaged in an activity. This state is associated with positive emotions and strong concentration.

Experiential marketing seeks to create flow experiences that deeply engage consumers, making their interaction with the brand rewarding and memorable. Immersive campaigns, such as interactive experiences or augmented reality settings, aim to achieve this state.

Traditional Marketing and Neuromarketing: some differences

The main differences between traditional marketing and neuromarketing concern the approach to the consumer (Kolar, 2014), the analysis techniques and the attention to unconscious reactions: while the former focuses on external and declarative methods, neuromarketing aims to measure unconscious physiological and psychological reactions, based on neuroscientific evidence. Similarly, while traditional marketing is based on consolidated techniques of communication and market segmentation, neuromarketing uses neuroscience to explore the instinctive and involuntary processes that guide purchasing decisions, focusing on deeper and less visible aspects of the human psyche. It can therefore be said that neuromarketing represents in many ways an important evolution in the process of developing and designing more precise and optimized communication strategies, capable of responding more effectively to the emotions and deep desires of the consumer.

More specifically, it is possible to summarize the main differences between the two approaches, highlighting the following:

1) Objectives and Approach

Traditional Marketing: Focuses on creating compelling messages and building relationships with consumers through tools such as advertising, promotions, public relations and direct selling. The goal is primarily to attract and retain consumers, working with a limited understanding of internal mental processes.

Neuromarketing: Its goal is to understand how emotions, visceral reactions and unconscious mental processes influence purchasing decisions. The approach is scientific and based on data obtained from neuroscientific studies, such as the use of eye-tracking, fMRI and EEG, to measure biological and psychological reactions to marketing stimuli.

2) Focus on Rationality vs. Emotionality and the Unconscious

Traditional Marketing: It assumes that consumer decisions are driven by rational needs and desires, although it understands the importance of emotions. However, it tends to operate on a more superficial level, trying to create positive associations through explicit messages and symbols.

Neuromarketing: It aims to understand and stimulate the emotional and unconscious reactions of the consumer, based on the evidence that many decisions are driven by processes outside of awareness. Emotions and cognitive biases are therefore at the center of neuromarketing strategies.

3) Techniques and Methodologies

Traditional Marketing: It uses traditional market research techniques, such as surveys, focus groups and interviews. The data collected is based on declarative and voluntary responses, where consumers express their preferences and perceptions directly.

Neuromarketing: It uses neuroscientific tools such as fMRI (functional magnetic resonance imaging) to observe brain activity; EEG (electroencephalogram) to measure brain waves; Eye-tracking to monitor visual attention; Biometric measurements such as heart rate and skin conductance to evaluate emotional reactions. These techniques allow us to obtain more objective and precise information than traditional methods, revealing non-conscious reactions that consumers are unable to express verbally.

4) Reaction Time and Instant Responses vs. Reflective Responses

Traditional Marketing: Often relies on reflective, rational responses, assuming that consumers take time to process the message before acting. Decisions are seen as logically considered.

Neuromarketing: Analyzes immediate responses to marketing stimuli, based on the idea that many decisions are impulsive and happen in milliseconds. Techniques such as eye-tracking can capture these instantaneous responses, indicating which elements capture attention and interest more quickly and accurately.

5) Understanding Consumer Behavior

Traditional Marketing: Segments audiences based on demographic and psychographic factors (e.g., age, income, interests), seeking to understand the overall motivations behind purchasing behavior. This view of the consumer is often idealized and based on general patterns.

Neuromarketing: Segments consumers not only by demographics, but also by biological and psychological responses to stimuli. It uses personality theory and more detailed psychological profiles, allowing for the development of marketing messages tailored to specific emotional responses and cognitive biases.

CONSUMPTION AND NEUROMARKETING STRATEGIES

As discussed in the previous part, neuromarketing builds its theories and application strategies on the analysis of how the human brain reacts to specific stimuli, often in an unconscious or semi-conscious manner (Tucker, 2000).

The structure of the human brain is characterized by the presence of three main areas:

1. *The cerebral cortex*: responsible for higher cognitive functions, such as thinking, memory, imagination, and language.
2. *The limbic system*: considered the center responsible for emotions and motivation. It includes, among other structures, the amygdala (which regulates emotions such as fear and pleasure) and the hippocampus (which governs functions such as memory and learning).
3. *The brainstem*: which controls vital functions such as respiration and heartbeat.

Furthermore, the brain is divided into two hemispheres: the *left hemisphere*, which governs language and logical-analytical thinking, and the *right hemisphere*, associated with creativity, emotions, and imagination.

To complete this brief schematic overview, it is also essential to consider *neural networks* and the role of *neurons* (which transmit electrical and chemical signals) and *neurotransmitters* (molecules that regulate emotions and behaviors, such as the well-known dopamine and serotonin).

Moving forward in our analysis, we can state that, in relation to the external world, the senses (sight, hearing, touch, taste, and smell) capture stimuli from the surrounding environment and transmit them to the brain. The brain then converts these stimuli into electrical or chemical signals and processes them through the *sensory cortex*, the *limbic system*, and the *brainstem*. Subsequently, these external stimuli, once received and processed, trigger a complex organic response (in terms of attention, memory, emotions, and actions), which can be either *conscious* (a reasoned decision) or *unconscious* (impulsive and automatic).

One of the main objectives of neuromarketing is to analyze these complex interaction processes between the consumer's brain and various types of products to evaluate, in particular, how external stimuli (advertisements, packaging, messages) activate specific brain areas. To this end, neuromarketing focuses on studying the *amygdala* (to evoke intense emotions), the *hippocampus* (to enhance brand memorization), and the *prefrontal cortex* (to influence purchasing decisions).

This field of study relies on research based on the techniques outlined in the previous chapter, including:

- *Eye-tracking* (analyzing gaze patterns);
- *Electroencephalography (EEG)* (measuring brain activity);
- *Functional magnetic resonance imaging (fMRI)* (identifying which brain areas are activated);
- *Biometric measurements* (heart rate, perspiration, etc.).

As we have seen, in the recent history of cognitive neuroscience applied to marketing, multiple theoretical perspectives have emerged, with several influential scholars becoming key references in analyzing the relationship between reasoning-based behaviors and more instinctive, emotion-driven behaviors.

One of the most prominent and frequently cited authors in international literature on this topic is Antonio Damasio, a neuroscientist and author of works such as *Descartes' Error*, who has developed a groundbreaking theory on the relationship between reason, emotions, and feelings, fundamentally reshaping our understanding of the role emotions play in decision-making and consciousness[6].

The key points of his theory are as follows:

1. *Emotions as bodily phenomena*

According to Damasio, emotions are *physiological responses* generated by the brain in reaction to internal or external stimuli. These responses involve the *autonomic nervous system* and trigger bodily changes such as variations in heart rate, muscle tension, and hormone production. For example, fear can activate a "fight-or-flight" response, leading to an accelerated heartbeat and an adrenaline surge. These bodily responses are an integral part of emotions, which we experience even before becoming fully aware of them.

2. *Emotions and feelings*

Damasio makes a clear distinction between *emotions* and *feelings*. Emotions are *automatic and unconscious bodily reactions,* whereas feelings are the *conscious perception* of those emotions—the way our brain interprets and makes sense of the bodily signals produced by emotions. For instance, when we notice our heart beating faster due to fear, our mind processes this physical sensation, transforming it into a consciously experienced feeling of "fear." In this sense, *feelings act as a bridge between the body and the mind.*

3. *Emotional reasoning*

Western thought, beginning with Descartes, traditionally viewed reason and emotions as *separate and conflicting entities*. Damasio, however, demonstrates that emotions are *essential for rational decision-making*. Through his studies on patients with *damage to the prefrontal cortex,* he observed that individuals who had lost the ability to experience emotions struggled to make effective decisions, even though their logical and cognitive faculties remained intact. This finding highlights that *pure rationality is insufficient* because *emotions* help us assess what is important and anticipate the consequences of our actions.

A key concept in this regard is that of *somatic markers*, bodily signals linked to past experiences that help guide future decisions. These markers rely on *emotional memory* and allow us to make intuitive choices, shaping what we perceive as rational reactions.

4. *Consciousness*

Damasio's theory extends even further, arguing that feelings play a *crucial role in the formation of consciousness*. He distinguishes between two levels of consciousness:

- *Proto-consciousness*, which is linked to emotions and bodily regulation.
- *Extended consciousness*, which builds upon feelings and the ability to reflect on them, incorporating elements such as language and self-narration.

In this framework, *feelings provide the foundation for self-awareness*, which, in turn, underpins our capacity for rational thought.

Implications for Neuromarketing

In summary, Damasio challenges the traditional *dichotomy between reason and emotion*, demonstrating that they are *interdependent and indispensable* for human behavior and decision-making. According to his theory:

- Emotions are automatic and unconscious bodily reactions.
- Feelings represent the conscious perception of those emotions, bridging the body and mind.
- Reason cannot function without emotions—they provide an intuitive guide that enhances rationality.

This intricate connection between body, mind, and emotions is fundamental for understanding human consciousness and decision-making.

Building on Damasio's theoretical framework—further explored by many other scholars—neuromarketing leverages the relationship between emotions, feelings, and reason to analyze and influence consumer behavior.

- Emotions serve as an unconscious guide.
- Feelings translate emotions into conscious experiences.
- Reason integrates these processes, often making decisions that appear rational but, in reality, are profoundly shaped by emotional factors.

These insights have practical implications that can transform marketing into a more human-centered and psychologically informed process.

Practical Applications in Marketing

Here are some concrete examples of how this relationship between emotions, feelings, and reason can be applied in marketing:

1. Emotions as a guide for purchasing decisions

Example: A consumer enters a luxury store and is greeted by a pleasant fragrance, relaxing music, and warm lighting. These sensory stimuli unconsciously trigger positive emotions (pleasure, comfort, desire) that influence purchasing behavior. Even if, on a rational level, the customer considers the purchase unnecessary, somatic markers associated with positive emotions push them toward buying.

Neuromarketing researchers use techniques such as heart rate monitoring and perspiration analysis to assess the emotional impact of environmental stimuli on purchasing behavior.

2. *Conscious feelings and brand loyalty*

Example: A consumer repeatedly chooses a particular brand because they feel like they are part of a larger community (e.g., Apple users). This sense of belonging stems from past experiences with the brand and positive emotional associations (advertising, storytelling, appealing design).

Neuroscientists analyze brain activity to identify activation in reward-related areas (such as the nucleus accumbens) when consumers see a familiar and beloved logo

3. Fear and urgency as emotional triggers

Example: A limited time offer displayed with a banner reading "Only available today!" can create a sense of urgency or fear of missing out (FOMO). This activates the amygdala, the brain region responsible for fear-related emotions.

Techniques such as eye-tracking and EEG (electroencephalography) help researchers analyze how visual stimuli that evoke fear or urgency increase consumer attention and emotional activation.

4. *Emotional storytelling and rational engagement*

Example: Advertisements that use storytelling—such as a touching holiday commercial about a family reuniting—evoke empathy and emotional connection. Even if the product itself is not essential, consumers associate the brand with deep emotional values, making them more likely to purchase.

Functional MRI (fMRI) scans reveal how effective storytelling activates the limbic system (emotions) and the prefrontal cortex (rational evaluation), showing how emotions and reason work together.

5. Packaging and sensory design

Example: A chocolate brand designs luxurious packaging with warm colors and a velvety texture. This creates an emotionally pleasant experience even before consumption, reinforcing the expectation of satisfaction and guiding the purchasing decision.

Tests such as visual preference analysis and fMRI scans show how attractive packaging design influences consumer perception and preference.

6. Subliminal advertising messages

Example: Some ads include subliminal stimulus such as fleeting sounds or brief images—that trigger unconscious emotional responses without the consumer realizing it. These stimuli can subtly shape purchasing decisions.

Brain activity monitoring can help researchers identify which subliminal stimuli activate emotional brain regions (such as the amygdala) without engaging conscious reasoning.

As neuromarketing continues to advance, ethical and regulatory challenges have emerged, shaping its implementation across various industries. The ability to influence consumer behavior at a subconscious level raises significant concerns regarding manipulation, autonomy, and informed consent. Ethical considerations become particularly relevant as companies collect biometric and neurophysiological data, often without consumers fully comprehending the implications of such practices. Consequently, addressing these concerns is essential to ensure that neuromarketing remains a responsible and transparent discipline (Ulman et al., 2015).

One of the primary ethical dilemmas in neuromarketing lies in the distinction between persuasion and manipulation. While traditional advertising aims to shape consumer preferences through rational appeals and emotional engagement, neuromarketing delves deeper, directly targeting unconscious decision-making processes. This raises pertinent questions about whether consumers genuinely exercise free will in their purchasing decisions or if their choices are subtly engineered by external influences. The potential for manipulation becomes more pronounced when consumers are unaware of the techniques being employed, leading to ethical debates on the fairness and integrity of such practices.

Furthermore, privacy and data security are critical concerns in neuromarketing. As businesses increasingly collect and store sensitive neurological data, the potential risks associated with data breaches and unauthorized usage become significant. Unlike conventional consumer data, neurophysiological information is deeply personal and can reveal insights into an individual's cognitive processes, emotions, and even mental health conditions. The unauthorized exploitation of such data could lead to discriminatory practices or invasive marketing strategies that compromise consumer well-being. Hence, establishing robust data protection protocols is imperative to prevent misuse and uphold ethical standards in neuromarketing.

To address these ethical concerns, regulatory frameworks have been introduced in various jurisdictions. The European General Data Protection Regulation (GDPR) mandates that companies obtain explicit consent before collecting biometric and neurophysiological data. This regulation ensures that consumers are adequately informed about data collection practices and have the opportunity to opt out if they choose. Similarly, in the United States, the Federal Trade Commission (FTC) has issued guidelines on neuromarketing practices to protect consumer rights and ensure transparency. However, enforcement of these regulations remains a challenge, as technological advancements often outpace legislative developments. Policymakers must continuously adapt regulatory measures to keep pace with the evolving landscape of neuromarketing technologies (Hoofnagle et al., 2019).

Corporate social responsibility (CSR) also plays a crucial role in mitigating ethical concerns in neuromarketing. Companies that prioritize ethical neuromarketing practices and embrace transparency foster greater consumer trust and brand loyalty. Ethical brands openly disclose their use of neuromarketing techniques, provide clear explanations of their data collection methods, and emphasize consumer well-being. By doing so, businesses can align their marketing strategies with broader societal expectations for responsible corporate conduct. This approach not only benefits consumers but also enhances the credibility and sustainability of businesses operating in the neuromarketing domain (Yfantidou).

Additionally, ethical considerations extend to the impact of neuromarketing on vulnerable populations. Children, elderly individuals, and those with cognitive impairments may be particularly susceptible to neuromarketing influences, raising concerns about potential exploitation. For example, children may lack the cognitive maturity to recognize persuasive marketing techniques, making them more vulnerable to subconscious influences. Similarly, elderly consumers or individuals with cognitive impairments may struggle to critically assess marketing messages, increasing their susceptibility to manipulative advertising. In response, advocacy groups and policymakers have called for stricter regulations to protect these populations from undue influence and ensure ethical advertising practices.

In conclusion, while neuromarketing offers powerful tools for understanding consumer behavior and enhancing marketing effectiveness, its ethical implications cannot be overlooked. The delicate balance between leveraging neuroscience for commercial gains and safeguarding consumer autonomy remains a key challenge. Moving forward, a multifaceted approach combining regulatory oversight, corporate responsibility, and ethical research practices will be essential to ensure that neuromarketing serves both businesses and consumers in a fair and transparent manner. By fostering an ethical framework for neuromarketing, stakeholders can ensure that advancements in this field contribute to consumer empowerment rather than exploitation, ultimately promoting a more responsible and sustainable marketing ecosystem.

EXPLORATIVE ANALYSIS AND CASE STUDIES

Neuromarketing, among its many applications, acquires particular importance in the food sector, by virtue of the strong emIn 1975, the Pepsi Coca-Cola company decided to launch an event called the Pepsi challenge, which consisted of an experiment conducted in supermarkets around the world, in which stands were offered with anonymous glasses, one of Pepsi and the other of Coca-Cola, and consumers were then asked to express a preference. The results of this experiment revealed that more than half of the subjects preferred Pepsi to Coca-Cola, but this could not be explained because this did not happen in the real market, where Coca-Cola had a significantly higher market share than its competitor. A possible explanation, put forward by Malcolm Gladwell, was that in a slip test, that is, tasting only one drink, consumers tend to prefer the sweeter drink, namely Pepsi, but for larger quantities this is not valid. After some time, in 2003, the neuromarketing pioneer Read Montague carried out the same study again by analyzing the subjects' brain activity using fMRI (Klinčeková, 2016).

This time, however, he structured it in two parts: initially the glasses remained anonymous and the same result as the first test was recorded again, that is, that about half of the sample analyzed preferred Pepsi, and this was confirmed by the intense activity of the "ventral putamen", an area of the brain that is activated when a taste is found attractive.

The second part of the experiment was to tell the volunteers what drink it was: well, 75% declared that they preferred Coca-Cola. Furthermore, anatomically, if I note a change in brain activity as marked by the activation, in addition to the ventral putamen, of the median prefrontal cortex, a part of the brain that is the seat of discernment thought. This happened because there was a contrast between the emotional part and the rational part of the subjects, from which suggested the superiority of Coca-Cola. The trust in the brand, the advertising, the logo, the brand values, the television commercials of Coca-Cola had won over consumers, to the point of making it so familiar to them that they unconsciously preferred the taste too. Emotional connotation of food and wine products: in fact, food is able to stimulate all senses at the same time, generating an all-encompassing experience. Eating is not an experience linked exclusively to the intrinsic characteristics of food: it is conviviality, it is being able to savor new flavors, it is being able to take a break, it is psychological and physical well-being and is, therefore, an activity that various factors can influence. For this reason, in the purchasing process, the packaging, the smells, the presentation of the product and even the storytelling associated with it will also be decisive.

Acting on those elements that directly or indirectly influence a consumer's sympathy towards the brand, or the product can be a weapon to obtain the desired success on the market.

There are several Neuromarketing investigations in the food sector, as already observed in the Coca-Cola/Pepsi case.

Let us now focus our attention on Frito-Lay, a subsidiary of PepsiCo, which produces, distributes and sells chips and other types of snacks.

In 2008, the American company, leader in the food sector, relied on neuromarketing tools, in order to increase the revenues of its Frito-Lay brand, with particular emphasis on the sale of low-calorie snacks aimed at women. The brand, in fact, according to market results, was well aware that its products were often avoided by women, because, presumably, they were too rich in calories. And in fact, research had suggested that, in general, only 14% of American women allowed themselves a salty snack between meals, while 25% preferred sweet foods and the remaining 61% opted for fruit, vegetables and drinks (Srinivasa, 2022).

The image associated with salty snacks was, in short, not very positive, as it was associated with an unhealthy lifestyle.

But what was it, specifically, that conveyed this idea?

Neuromarketing research by the company NeuroFocus was able to highlight that the main problem lay in the packaging.

The bright yellow color of the bag of chips, in fact, caused the activation of the anterior cingulate cortex, linked to the feeling of guilt, greater in women than in men: in other words, seeing that product, consumers felt a certain reluctance to buy it, because they were convinced of the poor quality, in terms of health, of that snack, considered too high in calories.

These considerations led, therefore, to the need to create a new package, characterized by a brown-beige color, which recalled nature. Furthermore, information and images were inserted on the bag that allowed us to create the idea of a healthy food, such as the indication of the spices used, the olive oil used and the baking.

With these small but decisive measures, it emerged that the sense of guilt was decidedly lower and, indeed, the brand's turnover increased by 8% in the following year.

The case offered by Frito Lay has demonstrated that "neuromarketing is not a science that forces consumers to make certain choices but rather helps the various brands to better design and communicate their corporate messages, making them more convincing and captivating than the products themselves" (Diana Lucaci, president of the Neuromarketing Science and Business Association).

As we have already understood, neuromarketing research, since it is based on a scientific, rigorous and methodical approach, can be extended to various sectors.

Among these, the automotive sector was also the subject of an interesting study, commissioned in 2002 by Daimler Chrysler, a German manufacturer of cars and means of transport for military and civilian use, with headquarters in Stuttgart, Germany, now "Daimler AG" after the sale of the Chrysler group.

The purpose of the research was to analyze the reactions of some subjects in front of images of multiple cars, from sports cars to everyday ones, including Mini Cooper and Ferrari, in order to better understand what psychological factors were triggered in the mind of the consumer.

Using functional magnetic resonance imaging as an analysis technique, some interesting aspects emerged from the test: first of all, the image of the Mini Cooper in the participants caused the activation of a small area in the posterior part of the brain, the "fusiform facial area", which is usually associated with human faces and their recognition. In a certain sense, it was as if the subjects recognized a familiar face in the car.

The conclusion that the researchers reached was, therefore, that probably the great success of the Mini Cooper lay in its "familiar" design, every day, safe and reliable like a friend. And, among other things, more associated with a young person.

Later, the researchers showed images of high-performance sports cars, including Ferrari and BMW. Well, from the magnetic resonance, it turned out that in the subjects subjected to this test, those parts of the brain associated with the concepts of power, social status and wealth, as well as prestige and satisfaction, were activated (Hammou et al., 2013).

It was clear, therefore, that the design and structure of the car, together with the reputation of the brand, were factors of considerable impact on the idea that consumers had in their minds.

Furthermore, in 2003, Chrysler promoted a new experiment, subjecting an exclusively male audience to images of 66 cars, to analyze their brain activity once again, through fMRI. In this case, what was basically expected was obtained: about 66% of the subjects wanted sports cars, 25% limousines and the rest preferred compact cars.

What happens to the brain this time? In fact, in front of such images, the neural circuits of gratification and the search for justification for expensive purchases were particularly involved (Scanagatta, 2021).

"Reward" and "reinforcement": this was what was triggered in the minds of consumers, as highlighted by the psychiatrist and neuroscientist Hernik Walter, involved in the study.

In fact, the participants preferred those cars with a sportier line, with chrome finishes and lowered bodywork: unconsciously, in fact, they thought that such elements could be a way to assert themselves in society and attract the female sex.

The car, according to what has appeared, is not a simple means of transport, but also has a symbolic value: on the one hand it can be "familiarity" as in the case of the Mini Cooper, on the other hand wealth as for Ferraris, or even strength and vigor for the sports ones, business and prestige for the more voluminous ones. A direct consequence of this will be the need to model advertising based on the audience you want to reach, proposing different storytelling for different interlocutors.

One area in which neuromarketing research can certainly be of great benefit is advertising. In fact, through specific tools, such as EGG or eye-tracking, it is possible to analyze how consumers react to advertising stimuli, whether they are attracted or indifferent.

Recent studies have shown that it is generally easier for the brain to memorize very short advertisements of about 1.5 seconds, furthermore the areas that are activated by the vision of an advert are the midbrain and the ventral striatum, i.e. the pleasure centers of gratification. Obviously not all advertisements are able to affect us in the same way, in general it has been found that the most memorable advertisements are those that activate our logical area, i.e. that produce a fast electrical response in the left frontal lobe, since they require an effort to memorize them.

Furthermore, within an advert there may be some parts that involve us more and others less. Precisely for the purpose of measuring the impact of the public, in 2009 Microsoft marketing manager commissioned me to do neuromarketing research to two companies: EmSense and Mediabrands. In the research, volunteers were shown Kia Motors and Hyundai Motor commercials, but from two different advertising channels: TV commercials and an advertising campaign from Xbox Live.

The subjects were applied geometric instruments for the detection of brain activity, the data concerned five main parameters: preference, memorization, attention, purchase intention and emotional cognitive response. The sample was divided into two groups: the first group was shown commercials from Xbox Live, while the second group was shown TV commercials lasting about 30 seconds. Both then had to fill out a questionnaire on the pleasantness and memorization of the observed content.

The experiment showed that the emotional involvement and memorization of the volunteers who only watched the advertising via Xbox Live was much greater than the TV commercials. Microsoft therefore greatly benefited from this research as it was able to prove that it was more convenient to invest in the advertising channel rather than in another.

In 2008, managers of the American company Campbell SUP undertook neuromarketing research to understand the reason for their soup's mediocre performance, despite the growing sector. For this research, Robert Darwin, vice president of the company, relied on other neuromarketing companies: Innerscoe Research Inc., Merchant Mechanics and Olson Zaltman Associates. The objective was to examine how consumers perceived the product on the shelf if they appreciated the packaging.

The experiment involved 40 subjects and took place in two phases: initially traditional marketing research consisting of an interview at home and subsequently a simulation of a purchase at the supermarket, wearing AI trading tools and biosensors connected to the body.

During the first part of the research, it emerged that consumers did not particularly appreciate the soup label.

In the next phase, the neuromarketing phase, I noticed that volunteers were not attracted to the soup because there were many similar products on the shelves, they confused the packages because they had the same colors and finally the choice was made quickly in a routine way, without paying attention to the difference between soups. The company understood through this research that more engaging recognizable packaging was needed for Campbell's soup.

The images show the labels of the soup and how it was changed after the neuromarketing research: they tried to make the soup more inviting by giving it the appearance of a hot meal, eliminating the spoon and renewing the packaging with a more elegant font.

However, despite the new packaging not being much appreciated by consumers, the company did not see significant sales increases: the causes were probably Stefano's high price and poor quality of the soup.

REFERENCES

Adams, M. (2020). The kingdom of dogs: Understanding Pavlov's experiments as human–animal relationships. *Theory & Psychology*, *30*(1), 121–141. DOI: 10.1177/0959354319895597

Foxall, G. R. (1987). Radical behaviorism and consumer research theoretical promise and empirical problems. *International Journal of Research in Marketing*, *4*(2), 111–127. DOI: 10.1016/0167-8116(87)90003-6

Glaenzer, E. (2016). *Are the Brain and the Mind One?* Neuromarketing and How Consumers Make Decisions.

Hammou, K. A., Galib, M. H., & Melloul, J. (2013). The contributions of neuromarketing in marketing research. *Journal of Management Research*, *5*(4), 20. DOI: 10.5296/jmr.v5i4.4023

Harizi, A. (2021). Imputing Emotions to Foods.

Hoofnagle, C. J., Van Der Sloot, B., & Borgesius, F. Z. (2019). The European Union general data protection regulation: What it is and what it means. *Information & Communications Technology Law*, *28*(1), 65–98. DOI: 10.1080/13600834.2019.1573501

Kahneman, D., & Tversky, A. (2013). Prospect theory: An analysis of decision under risk. In *Handbook of the fundamentals of financial decision making: Part I* (pp. 99-127).

Kleinginna, P. R.Jr, & Kleinginna, A. M. (1988). Current trends toward convergence of the behavioristic, functional, and cognitive perspectives in experimental psychology. *The Psychological Record*, *38*(3), 369–392. DOI: 10.1007/BF03395030

Klinčeková, S. (2016). Neuromarketing–research and prediction of the future. *International Journal of Management Science and Business Administration*, *2*(2), 53–57.

Kolar, E. (2014). *Neuromarketing and marketing management: contributions of neuroscience for the traditional marketing mix* (Bachelor's thesis, University of Twente).

Morin, C. (2011). Neuromarketing: The new science of consumer behavior. *Society*, *48*(2), 131–135. DOI: 10.1007/s12115-010-9408-1

Nyoni, T., & Bonga, W. G. (2017). Neuromarketing: No brain, no gain!. *Dynamic Research Journals' Journal of Economics and Finance (DRJ-JEF)*, *2*(2), 17-29.

Rawnaque, F. S., Rahman, K. M., Anwar, S. F., Vaidyanathan, R., Chau, T., Sarker, F., & Mamun, K. A. A. (2020). Technological advancements and opportunities in Neuromarketing: A systematic review. *Brain Informatics*, *7*(1), 1–19. DOI: 10.1186/s40708-020-00109-x PMID: 32955675

Scanagatta, C. (2021). Exploring the consumer's brain: a neuromarketing approach.

Spence, C. (2019). Neuroscience-inspired design: From academic neuromarketing to commercially relevant research. *Organizational Research Methods*, *22*(1), 275–298. DOI: 10.1177/1094428116672003

Srinivasa, R. (2022). *Effectiveness of neuromarketing in food companies: hijacking brains* (Doctoral dissertation, Dublin, National College of Ireland).

Tucker, D. M., Derryberry, D., Luu, P., & Phan, K. L. (2000). Anatomy and physiology of human emotion: Vertical integration of brainstem, limbic, and cortical systems. *The neuropsychology of emotion, 14*, 56-79.

Ulman, Y. I., Cakar, T., & Yildiz, G. (2015). Ethical issues in neuromarketing:"I consume, therefore I am!". *Science and Engineering Ethics*, *21*(5), 1271–1284. DOI: 10.1007/s11948-014-9581-5 PMID: 25150848

Utriainen, T. (2020). Neuromarketing and consumer neuroscience–the evolution and current state of the Art, an integrative review.

Vecchiato, G., Cherubino, P., Trettel, A., & Babiloni, F. (2013). *Neuroelectrical brain imaging tools for the study of the efficacy of TV advertising stimuli and their application to neuromarketing* (Vol. 3). Springer. DOI: 10.1007/978-3-642-38064-8

Yfantidou, I. (2018). A neuromarketing perspective of green advertising. The influence of environmental advertising appeals and CSR to consumers. *Aristotle University of Thessaloniki, available at:*https://ikee. lib. auth. gr/record/297480/files/GRI-2018-21381. pdf*(accessed 26 February 2019).*

ENDNOTES

1 "NeuroMarathon 2022" (https://www.neuromarathon.world/edizione-2022).

2 Read Montague, *Perché l'hai fatto. Come prendiamo le nostre decisioni*, Raffaello Cortina, Milano 2008.

3 *Magnetic resonance imaging (MRI)* is a medical imaging technique used in radiology in to form pictures of the anatomy and the physiological processes inside the body. MRI scanners use strong magnetic fields, magnetic field gradients, and radio waves to generate images of the organs in the body.

4 https://www.engagebay.com/blog/dunkin-donuts-marketing-strategy/

5 Ulric Neisser, *Cognitive Psychology*, Psychology Press-Taylor & Francis Group, New York 2014.

6 Antonio Damasio, *Descartes' Error: Emotion, Reason and the Human Brain*, Vintage Books, London 2004.

Section 3
Unveiling Technological Revolution in the Current Industrial Paradigm

Chapter 5
From the Industry 4.0 Paradigm to the Industry 5.0: A Cultural Revolution Towards Sustainability

ABSTRACT

This chapter aims to open the path for future research in the literature on Industry 5.0 paradigm. Technology, as essential to achieving a competitive advantage, is present in every phase of the value chain, since all activities require technological skills in addition to market skills. Although the technological dimension is widespread in different areas, it is essential to identify an "internal sector" of the company in which these digital skills are generated. This sector can be represented by the technological development activity, which goes beyond the mere research and development activity, including the set of activities and knowledge aimed at improving and innovating products and processes. Consequently, the different industrial paradigms that follow one another highlight the need for knowledge of these innovative technologies as AI and how the human being must be placed at the center of each business process, in fact this is the essential point of the Industry 5.0 paradigm.

DOI: 10.4018/979-8-3373-3466-0.ch005

INDUSTRY 4.0 PARADIGM

Thanks to the "WORLD WIDE WEB" we have the opportunity to find information of all types. The accessibility of digital platforms has made traditional information retrieval methods such as libraries obsolete.

Technological change has led organizations to change the communication strategies of their product or service. There is a continuous dialogue of the crowd that listens, participates, and interacts in the development processes of organizations. Consumers have changed their behaviours due to the entry of new means of communication that make the purchasing processes not predictable and difficult to analyze. On the business side, the web is increasingly acquiring a leading role in strategic planning, moving to the constant use of the "digital" elements and less and less to the traditional ones of marketing. Nowadays, customers who decide to buy a new scooter use the Internet as their first consultation step; a little less than twenty years ago maybe you would buy a reference newspaper or go directly to your dealership.

From this point of view, it should be noted that the web must not be observed only as a new expression of electronic commerce (e-commerce), but as an instrument of influence in the time before the purchase by the consumer.

If you think of Web 4.0, information control and new interfaces (e.g. home automation) can represent new frontiers for the real modification of everything around us. The twenty-first century sees rapid technological advancement. This development of the information technology-technological segment encompasses many economic fields, first of all, the industrial field. Starting from the 10s of the 2000s, the concept of "industry 4.0" was established (or "industry 4.0"). this term indicates a renewal plan that sees industrial automation integrated with some production-type microtechnologies as the protagonist for the sole purpose of improving both the quality of production and, consequently, the work performed, and the amount of the aforementioned.

This definition of "industry 4.0" owes its name to the German physicists Henning Kagermann, Wolf-Dieter Lukas and Wolfgang Wahlster who used it for the first time at a conference held in Hanover in 2011, referring to the German industrial plan that always in the wake of this process of technological renewal, provided for investments in infrastructures, schools, energy systems, research institutions and companies to make Germany's production system more efficient, as well as more modern, and, finally, to bring the German manufacturing industry back to the highest global levels. From the date of that conference, the results in the German manufacturing field are evident and for this reason, many of the European countries (including England and France, which have a higher rate of technological renewal than others) to pursue this economic policy. Among the most important studies carried out in the field of industrial automation are those of McKinsey and Boston Consulting;

Italy will also be in the running with the observations made by the Observatories of the Politecnico di Milano.

These studies not only verify the complete efficiency in the production sector of this technological renewal but lead to defining the impact that industrial automation will have on the social and economic context. For this reason, some theorists have defined this process with the name "Fourth Industrial Revolution". From a study by the Boston Consulting, it emerged that the so-called "Fourth Industrial Revolution" focuses on some important points, first of all on the adoption of some enabling technologies; many of them are of recent origin but already present and that have never satisfied the requirements in the field of applied research or in the real field of production. Nowadays, however, the world economic market scenario is in continuous evolution and this leads to mass customization, a topic of main importance for some industrial branches, especially the manufacturing type (such as the automotive sector). Although subsequently compared to other OECD countries (Organization for Economic Cooperation and Development), Italy to has begun to appear on this new window of the world of the economy such as Industry 4.0.

In fact, the process concerning Industry 4.0 starts in Germany in 2006 with the inauguration of the "High-Tech Strategy" and aims to coordinate and finance the national players involved in the development of innovative technologies.

A few years later, the plan was revived under the name "High-Tech Strategy 2020". Providing a generic definition, it is possible to state in the first thesis that the term Industry 4.0 (or Industry 4.0) indicates the aptitude and propensity of industrial automation to some new production technologies in order to optimize working conditions and increase productivity and product quality of the plants.

It is inevitable that the change of course towards an approach different from the one taken as a reference up to now, determines effects on different levels. For example, on an institutional level, the fact-finding survey launched in February 2016 by the Chamber of Deputies Commission for Productive Activities, Commerce and Tourism were particularly relevant, with "the aim of contributing to the definition of an Italian Industry 4.0 strategy" .

The paradox lies in the fact that unlike what has happened in other countries, (such as Great Britain where the government presented in 2015 the document "Strengthening the UK manufacturing supply chains. An action plan for government and industry", which identifies six priorities to accompany and facilitate the transition to Industry 4.0: innovation, skills, access to finance, build capabilities in small and medium-sized enterprises, build greater collaboration between supply chains, create more resilient supply chains), it is not, at least for now, the preparation of a real programmatic and scenario document by the Italian government. Like a matryoshka, it is possible to imagine that the concept of Industry 4.0 also includes that of Manufacturing 4.0 or even a digital factory and intelligent industry. Reference is made to a new production model, based on the development of digital technologies, on the interconnection between different production components and on an approach oriented towards process and product innovation. The goal is to strengthen the production system of the more consolidated economies, trying to reverse the post-industrialization and outsourcing processes that have occurred in recent decades. As stated by Rullani (1999), a scholar already mentioned previously when attention was paid to the sources of innovation (see chapter I par.1), an economic process cannot affect all industrial production in the same way. It is, therefore, necessary to make a distinction between replicative or standard manufacturing, which is taken into consideration by the so-called "low-cost" countries, and innovative manufacturing which, on the other hand, can guarantee growing market shares.

The Industry 4.0 model focuses on this second type of manufacturing (i.e. the innovative one), to try to make it more efficient, thus relaunching its international competitiveness. In summary, the expression manufacturing 4.0 does not refer to the simple automation of production processes or to the growing capacity of calculation and data collection, both elements that have already been present for some time within our economies (in this case we would speak of Industry 3.0), but to its being "smart", that is, to that ability to constantly interact and make the circulation of information possible, made possible thanks to interconnected platforms on the Internet. Manufacturing 4.0, in fact, integrates the new enabling technological resources (which will be discussed shortly) in the different levels of the decision-making process by exploiting the innumerable benefits of the network. Boston Consulting (considered among the "Big Three" in the world of management consulting), after careful studies, stated that the "core" activity of the fourth industrial revolution concerns a particular form of technologies defined as enabling.

Some are not unprecedented; they are concepts already present but which have almost never been taken into consideration for the development of a business process.

Italy and Industry 4.0

As already mentioned at the beginning of this chapter, Italy has embarked on the road towards the fourth industrial revolution later than the other OECD countries. In 2016, the Calenda plan was introduced, which however did not achieve the expected results, since if we take as a reference the total value of investments in tangible and intangible assets connected to 4.0 technologies, equal to approximately 13 billion euros, the figure is positive. However, if it is compared with the number of companies benefiting, about 53 thousand, to which must be added those that have benefited from the hyper-amortization, it is easy to think that there is a possibility of expanding the range of action of the companies that adhere to the fourth industrial revolution.

It also emerged that 2/3 of the incentives went to medium-large enterprises; investments almost always concerned the machinery component (10 billion investments in tangible assets against 3 billion in intangible assets).

Thus, Stefano Patuanelli, Minister of Economic Development, on 18 December 2019 underlined these difficulties through a letter to "Il Sole 24 ore": "All this has prompted us to review some mechanisms and characteristics of the 4.0 world. First of all, we have given the measures greater stability by planning the revision in a multi-year perspective, so as to guarantee companies a medium / long-term breath. Above all, we have characterized the Transition Plan 4.0 with greater attention to innovation and green investments ".

Technological growth is a process that must also be supported with support in terms of training and information. For this reason, we have launched the measure of innovation managers and we want to create a solid and stable connection between the world of production and that of research.

The aim is to ensure an adequate level of technology transfer: shortly we will present the Atlante 4.0 project, the first portal created with the collaboration of Unioncamere to raise awareness of the structures that operate in support of the processes of technology transfer and digital transformation of companies; with Enea, we are drawing up a plan which, thanks also to the support of the National Innovation Fund, will contribute to the growth of investments in innovation in our country. The Transition is a great challenge, but our entrepreneurial fabric will be able to take it up by having the MiSE as its first ally.

It is therefore correct to imagine that in a short time there will be a change of course in the perspective of Italian companies, which will increasingly begin to become familiar with the fourth industrial revolution and with the peculiarities and distinctive elements that this brings with it.

In the wake of the "Fourth Industrial Revolution", Italy, which has repeatedly demonstrated its willingness to participate in it, has adopted an inclusive political strategy that is attentive to sustainability with the "4.0 Transaction Plan". It provides for a greater focus on environmental sustainability, technological innovation and also towards an innovative aesthetic and design renewal process. The main economic strategies used by the Italian industries active on the technological renewal front are:

- tax credit for investments in capital goods in order to incentivize and help industries that intend to invest in new and functional capital goods for technological transformation;

- tax credit for research, development, innovation and design, in order to take advantage of private spending on Research and Development to keep up with the various technological renewal processes of other companies;
- 4.0 training tax credit, to use investments in staff training on the new technologies to be used.

The concept of industry 4.0, as specified, mainly concerns manufacturing industries. When the main branch of the manufacturing sector, that is the automotive sector, meets the concept of industry 4.0, production technologies are integrated for the sole purpose of improving production, in terms of quality and in terms of quantity.

It is possible to identify four different types of industry 4.0 intervention areas within the field of automotive production:

- Interconnection: the direct ability to connect sensors, machinery or various devices to people via wireless technology to significantly increase the effectiveness of the product;
- Transparency of information: industry 4.0 provides a large amount of information useful to the operator in certain cases of important choices, always providing transparency. The process allows the operator in question to collect a large number of useful data and information
- Decentralized decisions: the interconnection and transparency of information allow operators to make decisions both inside and outside the production plants. This ability to combine local information and at the same time global helps drive better decision-making and increase overall productivity.

Technical assistance: industry 4.0 shifts the role of a man from a machine operator to one who makes decisions and solves problems. Assistance systems 4.0 are designed for support operators who have to make complex decisions to solve problems that are often urgent and at short notice.

Industry 4.0 also enters directly into the field of customization of the final product in the case of the manufacturing industry. The automotive manufacturer who, therefore, works under 4.0-type paradigms, manages to customize individual cars in a more elementary way, without losing efficiency. This is defined as mass customization and lies in the ability of an industry to provide the final customized product to a much wider market, fully exploiting the potential of technological development. With what has therefore been defined as the "fourth industrial revolution", the production processes will essentially focus on enabling technologies used for other software programs, such as the Cloud (storage space, processing or transmission of data saved on demand and, therefore, not only locally), the Internet of Things (better known as "IoT" and consists of a series of computerized devices connected to a network), Big Data and Analytics or cybersecurity (series of technological measures in order to safeguard sensitive user data). These technologies will make it possible to combine the various needs of the market without the final product losing quality. In the wake of this technological growth, Italian companies are also trying to make room for themselves and adopt a leading role. Italian companies have the opportunity to modernize, strengthen and integrate the supply network to which they belong and to redesign the processes productive and organizational on the basis of a paradigm that allows them to improve efficiency and innovativeness and therefore to stimulate demand and bring back to the national territory productions lost in the past (reshoring).

Despite the intentions of the various Italian companies in this regard, between the years 2017 and 2018, the Ministry of Economic Development collected and analyzed data regarding the diffusion of technologies among the various companies: on the total population participating in a direct way to the economic life of Italian industries, only 8.4% of companies use at least one of the innovative technologies resulting from the fourth industrial revolution; while companies that don't consider, don't use and don't even plan to use one of the aforementioned technologies reach 86.9%. Obviously, it is necessary to contextualize these data based on the size, economic, physical and geographical, of the various companies in Italy. Only 47% of large companies have exploited part of their own proceeds to technologically renew the company. In September 2016[eu1.1], the Italian government presented an economic-technological plan in order to help national companies to win the innovation 4.0 challenge. The Calenda Plan so-called) aimed, in fact, to provide companies with main resources, theoretical references and economic allocations and was made up of two important principles: skills and innovative investments. For skills, the aim was to increase education around advanced technologies, starting from school to universities, from technical institutes to doctoral courses in order to create a 360-degree culture and directly train professionals in the sector.

The first maxim, on the other hand, had the purpose of encouraging, through various instruments, investments from the economic and then technological points of view of the various national companies. Immediately after the introduction of this Plan, data was collected starting from the first half of 2017; they indicate an increase in orders for capital goods with peaks of + 11.6% for machinery and other appliances and expectations on future orders at the highest levels since 2010. These data are comforting as they allow us to frame Italy as one of the powers that could, albeit with a secondary role, be part of the great panorama of the fourth revolution, but in addition to these data it is necessary to consider the variable of which companies actually invest, currently, in industry 4.0 and how the tools provided have been used, in the practical sense from Piano Calenda.

Industry 4.0 Main Technologies

It is necessary to dedicate an in-depth study to the main technologies of Industry 4.0. First of all, it is necessary to start with artificial intelligence (AI), born already in the previous revolution, and which is playing a central role in Industry 4.0.

In AI, machine learning systems are implemented, and neural networks and advanced algorithms are used to analyze large amounts of data, automate processes, and improve decision-making. In particular:

1) Intelligent production: Production optimization, better efficiency, and cost reduction. Intelligent manufacturing systems can dynamically adapt to changes in demand, anticipate machine failures and optimize resource management. 2) Predictive maintenance: Constant analysis of data from machine sensors to predict when a certain piece of equipment will need maintenance. This allows you to plan corrective interventions before failures occur, reducing downtime and maintenance costs.
3) Quality and process control: Real-time monitoring of production processes, detecting anomalies and ensuring product quality. Computer vision systems and machine learning algorithms can identify defects or deviations from expected parameters.
4) Intelligent Supply Chain and Logistics: Optimizing supply chain management, helping to forecast demand, optimize inventories, improve product traceability and optimize transportation routes.
5) Digital Factories: Enable the creation of digital factories, where production processes are highly automated, interconnected and adaptable. Intelligent systems enable more flexible and customized scale production.
6) Autonomous systems and advanced robotics: AI is critical to building autonomous systems, such as industrial robots and self-driving vehicles, that can operate safely and collaboratively in industrial environments.
7) Advanced data analytics: AI analyzes large amounts of data from different sources, helping to identify patterns, trends and insights useful for making strategic decisions.
8) Machine learning for customization: In Industry 4.0, AI can be used to adapt production processes more precisely to individual customer needs, enabling more personalized and flexible production.

In recent times, however, thanks to the interaction and collaboration between systems, the image of the global market are changing, giving greater importance to mass customization, becoming of interest to the entire manufacturing sector.

The 9 enabling technologies defined by Boston Consulting are:

- Advanced manufacturing solution;
- Additive manufacturing;
- Augmented reality;
- Simulation;
- Horizontal and vertical integration;
- Industrial internet;
- Cloud;
- Cyber security;
- Big Data Analytics.

Internet of Things

The Internet of Things is one of the enabling technologies of Industry 4.0. The basic idea of this concept is the pervasive presence around us of a variety of things or objects, such as radio frequency identification (RFID) tags, sensors, actuators, or cell phones which, through unique addressing schemes, are able to interact with each other and collaborate with their neighbours to achieve common goals. IoT is, for example, a refrigerator that orders milk when it "notices" that it is finished or it is a house that turns on the heating as soon as it hears you arrive. In this way we try, through this technology, to unite the real world with the virtual one. Undoubtedly, the main strength of the IoT idea is the great impact it has on different aspects of daily life and on the behaviour of potential users. Likewise, from the point of view of business users, the most evident consequences are equally visible in sectors such as automation and industrial production, logistics, business process management, and intelligent transport of people and goods.

Furthermore, the monitoring, control and transfer of information is of fundamental importance in order to be able to move accordingly. Consider in this case, as will be analyzed later, how this technology can be decisive for very complex sectors such as healthcare. Gartner Inc. (a multinational corporation that deals with strategic consulting, research and analysis in the field of information technology), expects that the IoT will reach 26 billion units by 2020, compared to 0.9 billion in 2009.

It is assumed that companies will invest in the IoT to redefine workflows in the factory, make the monitoring and control phases of materials more efficient and optimize distribution costs. To date, it is possible to list 5 types of Internet of Things used for the implementation of products and services:

1. Radio frequency identification (RFID): through this it is possible to automatically receive and identify data by radio waves, placing them in a tag and in a reader. The latter can store more data than traditional barcodes. There are three types of tags. Passive RFID tags rely on the radio frequency energy transferred

from the reader to the tag to power the tag. An application of this type can find space in the Advanced Mobile Location (AML) which allows a smartphone to recognize a call to an emergency number, activate the GPS autonomously, and automatically send the geographical coordinates to the emergency services via SMS with an accuracy of a few meters.

Active RFID tags may contain external sensors to monitor temperature, pressure, chemicals and other conditions. Semi-passive RFID tags use batteries to power the microchip as they communicate by drawing power from the reader.

2. Wireless Sensor Networks (WSN): Wireless Sensor Networks (WSN) consist of devices with autonomous sensors distributed in a space to monitor physical or environmental conditions. In addition, they can coordinate with RFID systems to better track the state of things such as their position, temperature and movements (as in the introductory example of radiators).

One of the technologies that make the most of this type of IoT is M2M (Machine to Machine), and indicates the technologies and services that allow the automatic transfer of information from machine to machine with limited or no human interaction. Examples of M2M are warehouse management equipment (which records incoming and outgoing material).

3. Middleware: Software that provides applications, services and capabilities, including data management, application services, messaging, and authentication, and that helps developers build applications more efficiently. It therefore acts as a connective between applications, data and users.
4. Cloud computing: the functions of this type of IoT are manifold. It can in fact ensure that the supplier makes available an entire program that the customer can use freely. Typical examples are email clients (such as Gmail).
5. IoT applications: Enable device-to-device and human-to-device interactions in a reliable and robust way. For example, FedEx through the use of this application monitors the temperature and position of its goods.

Cloud Computing

Cloud computing is a phenomenon that greatly characterizes the entire Industry 4.0 system. This technology allows the user to use it without ever owning it. It is based on the use of services and resources (such as storage, applications, programs, services) with the pay-as-you-go formula. This is the great revolution of this technology, as it allows you to create a new business model that is based on the reduction of internal IT power, to be able to acquire it externally as needed. Therefore, the user can use hardware and software at any time and place and from any device by accessing remotely without bothering to install, update, or backup the program.

The dematerialization of information represents a breakthrough for the new company. Precisely for the latter, everything translates into:

- More comfortable management;
- "Live" traceability of contents;
- Speed up access to documents;
- Efficiency in archiving.

An example that can concern anyone is electronic invoicing, or a digital process that leads to the transmission and storage of invoices in electronic format, completely abandoning paper.

In general, in the context of Industry 4.0 and artificial intelligence, it is possible to differentiate 3 types of Cloud Computing:

1. *SaaS (Software as a Service)* = it concerns the rental of an application that is in a system other than the one owned. Think of Google Scholar making it possible to search for academic articles.
2. *IaaS (Infrastructure as a Service)* = for sure it can be considered as the most useful and the most requested by companies. The latter, instead of owning resources such as licenses, patents or software, can take advantage of these resources via the cloud.
3. *PaaS (Platform-as-a-service)* = in this case, the goal is to offer companies a platform in which there are tools and features necessary to make the distribution of applications more usable in a secure way. Only the largest technology companies in the world can promote the provision of PaaS solutions (such as Google App Engine, Oracle Cloud Platform).

Gartner Inc., (a multinational joint stock company that deals with strategic consulting, research and analysis in the field of information technology) carried out an analysis concerning the Cloud Computing phenomenon in Industry 4.0. The development of this technology has caused the distinction of 3 different figures with respective different tasks:

- Hosting Service Provider: provides services such as virtual, application and management servers following the pay-for-use model typical of Cloud technology.

- Cloud Broker: is the one who manages the cloud services, regarding their use, configuration and supply. It is also the intermediary between the Cloud Provider and the Cloud Consumer.
- Cloud Consumer: is the individual who uses the service provided by the Cloud Provider.

Augmented Reality

Before giving a definition of augmented reality, it is important to state that this concept is the other side of another paradigm, virtual reality (VR). The latter indicates a three-dimensional digital environment generated by the computer, which can be explored by a person, who at that moment becomes part of the virtual world and can perform a series of actions in it. However, they differ in the sense of immersion they are capable of offering.

In fact, they can be defined as 2 different approaches to use and interaction with content.

Virtual reality, in fact, through special VR viewers, transports the individual to a parallel place that can be explored at 360 °. In this way you lose contact with reality, perceiving only what is transmitted in a specific virtual representation.

The augmented reality approach (AR) is in some ways different. In this case we do not lose contact with reality, but this is in a certain sense "augmented" by superimposing texts, images, or representations of various kinds (an engineering professor can explain to his students the structure of the engine of an airplane with the the use of this technology. Pupils will then be able to "really" observe the engine of the plane while perceiving what they see in real time).

According to various critics and lovers of this phenomenon, it is possible to distinguish 2 different types of augmented reality (AR):

1. AR managed by smartphones, tablets, iPhones and IPads: widespread in particular in 2016 with the launch of "Pokémon GO", a video game designed for smartphones that allows users to train and capture Pokémon (cartoon present in the television scene at the beginning of the millennium). These virtual creatures that appear on the mobile phone display are superimposed on the images presented in real time by the camera and, therefore, are perceived by the user as if they were inserted into the surrounding context, as they move. This type of technology has the advantage of being particularly accessible, since it only requires the use of a smartphone and an ad hoc application: just frame the surrounding environment with the mobile phone camera and you can access all the contents provided by the software.

Being totally revolutionary, this new technology brings with it not only benefits but also various criticisms such as the one made by CODACONS (Coordination of associations for the defense of the environment and the rights of users and consumers), which affirms that Pokemon Go can be a concrete danger for those who use this app that distracts attention from the dangers encountered when in public places, presented a complaint to the Rome Public Prosecutor's Office, requesting the total ban on the spread of the Pokemon app Go on the Italian territory ".

2. AR managed by wearable devices (viewers, glasses): they allow you to enhance the experience and make it even more immersive, but have a higher cost than the first type of technology. Until recently, Google with its Google Glass monopolized the market for these devices. To date, Samsung is trying to patent contact lenses (Gear Blink) that allow individual to immerse themselves in augmented reality.

Big Data

This phenomenon has as its main and fundamental characteristic that of storing, managing and analyzing information, often of a personal nature (personal data or user data), produced by numerous sources, managed automatically through processors and algorithms (so-called machine learning) and filtered by means of an extraction process (so-called knowledge discovery in the database) that makes use of specific software programs (data mining and text mining or knowledge discovery in texts).

From a message via Whatsapp, to a photo, a Google search or a purchase in a store, all of these are examples of data. Almost every action that takes place today can be collected, analyzed and monetized. According to the 2010 "How Much Information" report, it seems that each of us generates an average of 12 gigabytes of data every day, and the number is constantly growing.

Until the early 2000s, Big Data was defined through three concepts: volume, speed and variety. Over the years, while the phenomenon became more and more concrete and applicable in companies, the model was modified by adding two more Vs (Ottolia, 2017). Here are some characteristics of Big Data:

1. *Volume:* refers to the amount of information that cannot be stored with traditional technologies. Given that the numbers of information grow year by year, it is difficult to anchor to a number, the limit value above which we can speak of Big Data.
2. *Speed:* the innovative technology of Big Data makes it possible to receive as much data as possible in less time. Just think of those devices that are able to collect data in real time.

3. *Variety:* in 2008 Chris Evans, an American journalist, wrote: << More isn't just more. More is different >>. In this case, reference is made to the multiple types of data that can be received and coming from heterogeneous sources. The main distinction that must be made is between:
    - Structured data = stored in databases, organized according to rigid tables and schemes. This is the most suitable type of data for relational information management models.
    - Unstructured data = held without resorting to any fixed scheme.
    - Semi-structured data = represent a synthesis between the former and the latter. Semi-structured data can be understood, for example those in HTML format. In this case there are no tables, but the language is still interpretable and classifiable.
4. *Truthfulness:* those who carry out their business having Big Data as their fulcrum will surely know the famous phrase "Bad data is worse than no data". The data must be a sure source, must bring out the objective truth. The problem is that the efficiency of this variable is offended by the first 2 (Volume and Speed), given that the large number and speed with which the data are received, make those who analyze them always be skeptical.
5. *Variability:* over the years there has been more and more to deal with a lot of data in different formats and from different contexts. The changeability of their meaning is an aspect to be taken into consideration when the data are received and interpreted. Even more so, if it is a user who manages a line of business and not just the data scientist.
6. Varieties: in recent years they have been defined as the new oil or the new gold, which is an inestimable source of value. Through the use of what are the tools of Analytics, we also consider the value that a company can extract from the huge phenomenon of Big Data.

A Focus on Some "Complementary Enabling Technologies"

So far, the enabling technologies in Industry 4.0 have been analyzed. Furthermore, it is correct to pay particular attention to another type of technology which, like those already examined, expresses different essential features of the fourth industrial revolution.

They take the name of "complementary enabling technologies", since they are already widespread in many companies, but they fall within the macro-area of enabling technologies only if used in a functional way (such as an e-commerce system that uses data deriving from Big Data).

Eight different types of complementary enabling technologies can be distinguished in:

1. Digital technological solutions of the supply chain aimed at optimizing the management of the supply chain and the management of relations with the various players;
2. Software, platforms and digital applications to coordinate and manage the logistics area by integrating activities (for example, SCM (Supply Chain management), CRM (Customer Relationship Management);
3. New online payment systems;
4. Geolocation system;
5. New technologies for the in-store customer experience;
6. EDI systems (Eletronic Data Interchange);
7. New e-commerce channels;
8. Creation of integration systems applied to the automation of business processes.

Geolocalization

As previously mentioned, this particular form of technology falls within the branch of "enabling" only if used in a functional way.

Precisely the geolocation system (or RTLS - Real Time Location System) is a declination of the IoT, since through the use of sensors, it is plausible to uniquely identify any object or individual (for example through a dashboard of a truck able to transport goods, it is possible to monitor the location and / or movement of the goods).

The main objective of the geolocation system is to make it possible to trace and detect valuable and strategic information to support management. For this reason, in Industry 4.0, it is taken into consideration in different application fields, for example, as will be seen later, in the health sector it is useful for monitoring bedridden or walking patients or to make the identification of patients more efficient. machinery and sanitary equipment.

In the traffic sector, this technology has caused a sharp decline in the use of taxis (paid public passenger transport service on the square), fueling the explosion of Uber (a company that provides private car transport service through a mobile application that connects passengers and drivers directly). The US company, to date, has not yet entered the Italian market, except for a branch of the company that has diversified its production lines, Uber Eats.

The types of geolocation are different, some more known and felt, others less. One of the most famous systems is the GPS which is able to define the position of an object, since this is connected to a satellite and communicates its spatial coordinates to it. Wi-Fi is also a type of geolocation, in fact, through the triangulation of multiple access points, it is possible to detect the position of an object.

Customer Relationship Management (CRM)

Before analyzing how systems and technologies dedicated to customer relationship management approach the world of the fourth industrial revolution, it is useful to illustrate what it actually means for companies to devote their entrepreneurial policies to CRM.

Until just before the 1990s, the only connection between customers and the company was the so-called switchboard, the complaints office or the employees who interacted with the outside world.

In that period begins the rise of a new business strategy that takes the name of customer relationship management and software that is able to answer any questions from customers.

In recent years, CRM systems have created a channel where companies can get to know the customer and their needs better, thus simplifying the decision regarding the strategy to be implemented. CRM systems learn customer information from various sources: emails, call centers, websites, shops, sales people, etc.

Figure 1. Information hub (Adapted from Geunes and Pardalos, 2003)

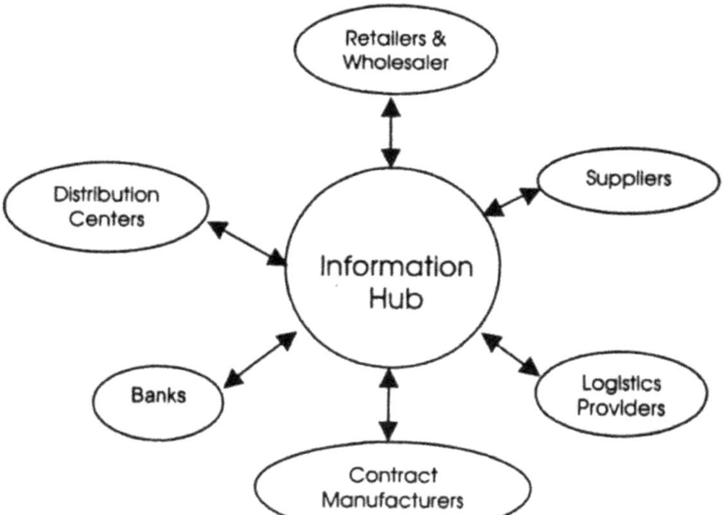

The output of this process is processed and analyzed by CRM software, now essential for companies wishing to compete in increasingly dynamic markets.

To define CRM, it is possible to say that the software is the main feature, but there is not only this. It must be observed and analyzed as a process, including activities towards the market and customers, at any stage of the chain. To ensure that a CRM strategy is successful, it is necessary to acquire and analyze market trends and customer information from multiple sources, in order to increase sales while leaving high customer satisfaction unchanged.

Today, digital transformation has changed almost the whole way of doing business: production, sales, communication and interaction with company stakeholders.

It was not only the company that changed, it must be said that the customer has also changed his way of approaching the market. The enabling technologies have greatly influenced the choice of the customer, since the latter with the spread of augmented reality has the opportunity to visit, via platforms, a cruise ship before booking his vacation.

To adapt to this new industry model, CRM must become an integral part of the digital transformation strategy and become a business process that affects customer behaviour, personalizing their experience with the company.

Here we could speak in this case of CRM 4.0, meaning this name as a platform that is able to connect processes, people and machines to create an innovative business model, the Smart Connected Factory, in which a new (intelligent) factory experience in which all the players interact with each other, exploiting know-how efficiently.

But how could the customer experience change for the customer? In retail until a few years ago there was talk of webrooming, or the possibility for the individual to observe the product online by anticipating the visit to the physical store, but without finalizing the sale. Today, however, we are starting to talk about "omnichannel", alluding to an increasingly clear integration between real and virtual. Real interactive showcases are created which, through enabling technologies (IoT and augmented reality above all), allow to enrich the user's experience with the company. Functionally with enabling technologies, the CRM paradigm fits into the new world of the fourth industrial revolution, guaranteeing efficiency at all points of the customer journey.

Supply Chain Management (SCM)

The management of the distribution chain is defined as the set of activities that aim to organize and make the flows of supplies and finished products more efficient.

Usually, several independent companies interact with each other in the manufacture of a product, delivering it to the last user in the chain who places it on the market. The subjects present within the chain are almost always: producers of raw materials, wholesalers, retailers and transport companies (La Londe and Masters, 1994).

Other theorists define supply chain management (SCM) as an alignment of companies that introduce products or services on the market. It is fair to note that this vision of SCM includes the final consumer as the last player in the chain.

Another definition observes that the SCM is a network of organizations involved through links upstream and downstream in the different processes and activities that produce value in the form of products and services provided to the final consumer (Christopher and Towill 2001). Therefore, trying to synthesize the different definitions promoted by scholars, the supply chain management is composed of upstream companies, downstream companies and the final consumer. Enclosed in this axiom, three degrees of supply chain complexity can be identified: 1) supply chain-direct, 2) supply chain-extended, 3) supply chain-ultimate.

As for the supply chain-direct, it is made up of a company, a supplier and a customer involved in the upstream and / or downstream flows of products, services or information.

The supply chain-extended, on the other hand, includes the suppliers of the first supplier and the customers of the first customer. All subjects are involved in both upstream and downstream flows as well as in the direct chain.

The supply chain-ultimate, unlike the first two, detects all the organizations involved from the supplier to the final customer. It shows precisely the complexity that this type of chain can achieve, since for example, a third supplier could provide financing, assuming part of the risk by offering financial advice. This briefly illustrates some of the many functions possible for complex supply chains.

After providing a fairly clear picture of the functions and types of supply chain management, in the next few lines we will try to analyze how the latter changes when it approaches the new Industry 4.0 model.

THE INDUSTRY 5.0 AS CULTURAL REVOLUTION

The Fourth Industrial Revolution has brought with it numerous social and ethical impacts that influence different aspects of society.

> On a social level, automation and the introduction of advanced technologies can lead to the reduction of jobs in some traditional industries, generating concerns about unemployment. Furthermore, to keep up with evolution, new skills are required and continuous training becomes essential to maintain relevance in the job market. It becomes a challenge for workers who must adapt to advanced technologies such as the Internet of Things (IoT), artificial intelligence (AI), and robotics. Precisely with respect to the aforementioned ability to adapt, and also regarding the accessibility of these technologies, economic disparities

between countries and within societies can widen. Nations and people with greater resources can more easily benefit from the opportunities offered by the fourth industrial revolution. Finally, the widespread use of sensors and connected devices can trigger privacy concerns. Massive data collection can lead to potential abuse and requires a balance between innovation and protection of personal data.

On an ethical level, with the increase in the use of AI-based systems and robots, ethical questions arise regarding the responsibility for actions performed by intelligent machines. "Who is responsible in case of errors or damage caused by automated systems?" it is only one of the first issues raised and a cause of general fear. Furthermore, ensuring equity in access to Industry 4.0 technologies is an important ethical concern. Ensure that technological innovations are distributed equitably to avoid social and economic disparities. The possible environmental impact is a widespread concern, which Industry 4.0 can greatly influence, both positively and negatively. On the one hand, advanced technologies can contribute to more efficient management of resources, but, on the other hand, the production of electronic devices and energy consumption can lead to new environmental challenges.

Finally, the increasing dependence on technology may raise ethical questions about our ability to manage a highly interconnected society dependent on automated systems. Public opinion: It is important to understand what public opinion is about Industry 4.0, to identify its problems, to provide information where criticism is reduced to bar chatter and to take steps together towards a more aware, productive and sustainable.

Past industrial revolutions have always given impetus and shape to change in society through the usage of new technologies and the consequent improvement in the quality of life, as already mentioned in the previous chapters. However, Society 5.0 requires a new paradigm for industry that goes beyond neoliberal capitalism. In this scenario, it is the industry that tries to adapt to society. Following in the footsteps of the United Nations Sustainable Development Goals, new concepts of social development, such as sustainability and human-centeredness, have become essential to corporate objectives. Consequently, industry is an essential component of society. Society's mobilization for civil rights and environmental conservation has increased. Solidarity towards Ukrainian citizens, the Black Lives Matter movement and the sounding board created around Greta Thunberg's speech have verified this change in society. The industry heeded the warning and realized the importance of refocusing not only on supply, but also on demand. She recognized that competition will focus not only on reducing costs, but also on challenges such as growing wealth and protecting the environment. Criticism, therefore, paints Industry 4.0 as an obsolete paradigm that can no longer respond to society's needs. It is neoliberal capitalism, founded on the concept of profit and the supremacy of shareholders, that is considered the cause of problems such as social inequality or the depletion of natural resources and environmental damage.

According to Nahavandi (2019), the paradigm that describes Industry 4.0 is not able to consider the human costs resulting from the optimization of production processes. The author, therefore, sees the lack of a social dimension in Industry 4.0 and indeed predicts that, if nothing changes in this sense, the benefits of Industry 4.0 will be completely canceled out by political and trade union pressure. To this end, it is necessary to implement a new model that capitalizes on the lessons and improvements made to overcome the pandemic crisis, together with a hyperconnected industrial ecosystem based on big data, which enriches and absorbs the ideals of Industry 4.0 and sustainable development.

Therefore, Industry 5.0 is the name given to the new paradigm. Unlike previous revolutions, Industry 5.0 recognizes the role it must play within society. The main objective for which it was conceived is to promote sustainable development, so that production processes respect the environment and the well-being of workers. Therefore, the concepts of Society 5.0 and Industry 5.0 are linked, not only in the sense of a change in processes or production methods, technologies and processes or internal organization of companies, but rather the start of a new new economic and social vision. In support of what has been said, the European Commission, through its studies, predicts a reduction in industrial polluting emissions by 25.1%. These figures are, however, contingent on the transformation of the economy, a change resulting from the adoption of the founding principles of Industry 5.0, including resilience and efficiency, sustainability, the principles of the regenerative and circular economy, as well as changes governance and policy and the adoption of Horizon Europe and national resilience and recovery programmes. In this sense, the vision of Industry 5.0 represents an economic model that is not only based on production and consumption, but which embraces the needs of Society 5.0.

The Fifth Industrial Revolution, therefore, focuses on a concept that puts human progress and well-being at the centre, based on the reduction and reorganization of consumption habits towards sustainable development. As already mentioned, Industry 5.0 is destined to act as a bridge in a transition aimed at overcoming the neoliberal capitalism model, which involves the reduction of the primacy of profit in favor of shareholders.

Below the three main pillars for Industry 5.0 intended as cultural revolution:

1) Human Centricity

It starts from the assumption that business processes must be designed by focusing on the role of man, his needs and his benefits. It is a human-centric revolution precisely because it has the task of bringing man back into production, freeing him from routine tasks. Consequently, from an operational point of view, the centrality of man requires that his function extends beyond the conventional management of processes to that of decision maker, assisted by technology. To this end, we recall the theme of "up-skilling" and "re-skilling", considered crucial in human-machine cooperation. Likewise, the current conception of Industry 5.0 sees the human touch in industry. It's about incorporating AI into human-supported operations in a way that amplifies the worker's capabilities. It is about technology at the service of man and not vice versa. This justifies the concept of Human Centrity. Furthermore, the current first pillar describes the transition from a progress whose driving force is technology to one in which society plays a fundamental role and therefore gives importance to human needs. It can therefore be stated that the fulcrum of Industry 5.0 is the balance that is generated between man, machine, values, tasks, knowledge and skills, which lead to personalized and individualized products. Finally, it is an Industry 5.0 priority to create a safe and inclusive working environment that promotes physical and mental well-being and acts as a protector of fundamental rights.

2) Sustainability.

In order to respect the boundaries of the planet, the industry must be sustainable. The task of Industry 5.0 is therefore to protect the environment and preserve natural resources while promoting economic growth. Also in this case, the industry, through the use of technologies such as AI and additive manufacturing, must aim to plan and define strategies that favor the development of technology based on keywords such as reuse, reuse and recycling of resources. Citing Bruntdland's 1987 studies, they propose the definition of sustainable development, according to which it is defined as such if it allows the needs of current society to be satisfied without affecting the capacity and therefore the resources of future society. Similarly, they discuss the importance of reverse logistics, circular economy and supply chains to achieve zero waste. Therefore, there are several points within the 'Sustainable Development Goals' plan defined by the United Nations that refer to the need for greater sustainability of the industry. Some aspects can be mentioned that relate to responsible consumption and production and industry, innovation and infrastructure.

3) Resilience.

This dimension refers to the challenges posed by the pandemic and the war in Ukraine and how society, but more particularly the industry, responded and learned from these shocks. Resilience is therefore defined as the typical ability of a system to resist or react quickly to certain exogenous factors that undermine or may undermine its correct functioning. Therefore, Industry 5.0 has the ability to manage uncertainties arising from markets, supply chains, customers and even entire national industrial systems, as well as geopolitical changes and natural emergencies. From a business perspective, the concept of resilience translates into greater agility and flexibility in responding to consumer needs. Production systems require the participation of consumers in the design phases to increase the possibility of customizing products and thus anticipate changes in market demand. Finally, robots serving workers are adopted to increase flexibility and minimize time to market.

The next section tries to illustrate case studies that can be conducted for exploratory purposes, aimed at improving the knowledge of a specific concept, and considered the prelude to research for explanatory or descriptive purposes used for casual investigations and the description of a given reality.

EXPLORATIVE ANALYSIS AND CASE STUDIES

AI Solutions in Novartis Pharmaceutical Company

Novartis (https://www.novartis.com/) is a Swiss pharmaceutical company that is important for the development of innovative products. The company is redefining medicine in order to improve and prolong people's lives. As a global leader in the pharmaceutical sector, we use innovative scientific and digital technologies to create transformative therapies in areas with significant medical needs. Committed to discovering new drugs, we are consistently among the top companies in the world in research and development investments. Novartis products reach over 750 million people globally, and we are working to find innovative ways to expand access to our latest treatments. Globally, approximately 108,000 people of 140 different nationalities work at Novartis.

Novartis is committed to becoming a leading pharmaceutical company, strong in advanced therapies and data science. The key strategy that helps Novartis is to maximize the use of data and digital technologies to achieve this goal. Data science and digital technologies allow us to reimagine how we innovate in R&D, interact with patients and customers, and increase operational efficiency. Novartis focuses its efforts on four priority digital strategic areas

- Implementing 12 flagship digital projects: building a strong foundation to launch our digital transformation
- Making Novartis digital: sharing, learning, and acquiring talent.
- Becoming the number 1 partner in the technology ecosystem: enriching Novartis with external expertise.
- Going beyond: with large-scale partnerships to address future, radical changes in health scenarios.

The pharmaceutical company in question and Microsoft have announced a collaboration to accelerate the research and development of new drugs, thanks to AI support. This agreement between the two giants aims to optimize how drugs are discovered and developed. The agreement between big Pharma and big Tech will also give rise to the Novartis AI Innovation Lab, a new laboratory that will seek to strengthen Novartis' expertise in artificial intelligence, from research activities to commercialization, accelerating the research and development of transformative drugs for patients worldwide. Initially, these joint research activities will focus on personalized therapies for macular degeneration, cellular and gene therapies, and drug design, and will take place in Basel at the Novartis campus, in Dublin at the Novartis Global Service Center, and in Cambridge at the Microsoft Research Lab.

The agreement envisions medicine through artificial intelligence and data science. The alliance between Microsoft and Novartis will support Novartis' commitment to using data and artificial intelligence to change, replace, and develop drugs as well as to commercialize them. Novartis will establish an AI Innovation Lab to allow its employees to use artificial intelligence in various activities. Joint research activities will include co-working environments and will focus on personalized therapies.

This multi-year agreement between the two giants will focus on AI Empowerment, where the Novartis lab will receive and enable the empowerment of AI through researchers, and AI Exploration, where Novartis will use AI to analyze and deliver intelligent therapies, starting from generative chemistry.

This agreement is a dream come true for science and technology enthusiasts, towards a future where every individual can effectively benefit from ongoing research efforts.

Roche and the Use of Big Data

Roche (https://www.roche.com/) has been alongside the health system for years to allow all citizens to receive treatment and guarantee the right to health today but also in the future. The pharmaceutical company Roche has 125 years of experience and has been engaged for years in initiatives aimed at supporting the value of real word data as a valuable source for answering research questions regarding clinical practices and healthcare organization. Roche is the only company in the health sector that embraces the patient journey at a genomic, clinical, economic, organizational and social level, service solutions, diagnostic tests and drugs

Roche Italia launched "Roche Now, big data united for health", a series of digital talks to understand how big data and artificial intelligence can improve public health and the health of citizens. It concerns:

- A journey into the world of big data applied to health, led by scientist Ennio Tasciotti, to analyze problems and opportunities with the System's experts.
- Personalized medicine, cutting-edge research and sustainability of the healthcare system are just some of the opportunities that more effective data management would bring to our country

Roche Now, big data united for health which is a guide for all citizens interested in understanding how the analysis of an enormous amount of health data can radically change the world of healthcare and improve the health of citizens: this revolution is already in place.

During the virtual event presenting the project "Roche Now, big data united for health", the first cycle of digital talks was launched conducted by the scientist Ennio Tasciotti which is developed through five video pills aimed at promoting greater awareness on issues such as Real World Data, Real World Evidence, FSE (Electronic Health Record), analyzing them from the point of view of the actors involved, from citizens to doctors, from institutions to technical experts in the sector and the media, with particular attention to current events. An efficient health data ecosystem has in fact often been cited as a key success factor in the response to COVID-19, but the pandemic has simultaneously made more evident the numerous barriers that still limit the use and sharing of data today. Several countries, as well as Italy, have limited capacity to collect, analyze and share high-quality data, resulting in negative impacts on the effectiveness and efficiency of the system. In the case of Italy, the extreme heterogeneity of approach at regional level represents a further limitation to the opportunities deriving from an integrated analysis of health data.

In the first Roche Now Talk, available on the official channels of Roche Italia, the world of big data is introduced and its terminology is explained; the second focuses on patients; the third features Vincenzo Valentini, Deputy Scientific Director of the Gemelli Polyclinic, who analyzes the potential of big data for doctors; the fourth appointment takes into consideration the world of institutions, represented by the former Minister of Health Beatrice Lorenzin; the fifth episode hosts the scientific director of the IIT (Italian Institute of Technology), Giorgio Metta, providing a technical perspective on the topic. Further discussion will follow, with other experts in the sector who will allow us to delve further and better understand the world of Big Data in healthcare.

The "Roche Now, big data united for health" project is sponsored by SIF (Italian Society of Pharmacology) and SIMM (Italian Society of Leadership and Management in Medicine) and during the round table also chaired by their representatives, opportunities emerged, but also structural and cultural critical issues that hinder the path towards the medicine of the future.

The combination of pharmaceutical and diagnostic expertise within the same organization has positioned Roche at the forefront of personalized medicine with a diverse portfolio of treatments in oncology, immunology, infectious diseases, and central nervous system. Research and the drive for innovation have always been the response offered by Roche to the continuous challenges of health, as evidenced by the incessant work carried out in the development of a unique portfolio in pathologies with a high degree of complexity, such as tumors, haemophilia, autism, Alzheimer's, spinal muscular atrophy (SMA) and multiple sclerosis.

For years this pharmaceutical giant has been at the top of the world rankings for investments in R&D. In the last year alone, Roche has invested 45 million euros in clinical research in Italy, with approximately 210 research centers involved in over 238 clinical studies and over 17,850 patients who have thus been able to benefit from a specific diagnostic path and treatment cutting-edge without any cost to families or the National Health Service. Always looking to the future of health, Roche has long been committed to the sustainability of the health system through numerous activities, convinced that in a social context characterized by an increase in average life expectancy and limited resources, sustainability means guaranteeing everyone the right to health, within a properly functioning system, without waste or flaws, and real equity of access.

Pfizer and Cloud Computing Support

The pharmaceutical company Pfizer (https://www.pfizer.com/) is now known to the general public for the European vaccine against COVID-19. The company was born in New York and started out with products such as Xanax, Viagra and Advil. Pfizer has addressed supply chain complications with doses of cloud computing, virtualization and other tools, allowing it to track shipments from a broad network of internal and external sources with greater efficiency and accuracy. Over the past year and a half, Pfizer has configured its systems for managing its supply chain in response to a variety of challenges. Such as the company's megamerger with Wyeth in 2009 brought its own integration challenges. More recently, the company's best-selling cholesterol drug, Lipitor, faces pricing pressure from new generic competition, forcing Pfizer to take steps to supply the pill as efficiently as possible. The consequences of supply chain problems fall on drug manufacturers and patients. Patients depend on drugs transported along supply chains to treat serious conditions, and drug makers risk losing revenue when shipments don't arrive, or problems arise in their supply chains. In some cases, pharmaceutical manufacturers are adopting new software and infrastructure to manage their supply chains with an eye to saving costs and improving regulatory compliance. Supply chain or logistics software that doesn't work properly can cause serious problems. Pfizer has chosen to leverage a cloud-based supply chain platform from GT Nexus as part of a comprehensive approach that allows everyone involved in the company's supply chain to connect to the system regardless of each site's IT infrastructure. And instead of having to adopt Pfizer's resource planning software for the company's 500 partners, all groups use the same information exchange framework that allows them to send and receive data from Pfizer.

Digital transformation has become essential in modern life, embracing every area from school to healthcare. Digital health has played a crucial role during the pandemic by offering assistance and treatment to patients, decreasing contagion, saving time and resources, guaranteeing a better quality of life. Without digital transformation we will have experienced this historical period with much more difficulty. Those who have invested in digitalisation will continue to grow for many more years, developing the new digital capabilities acquired during the health emergency. Italy should take advantage of this flourishing moment to grow on the market through digitalisation, thus being able to compete with other European countries. As regards the pharmaceutical sector, it is undergoing a fourth industrial revolution by adopting innovations such as Cloud computing, big data and many others, capable of making these industries more competitive on the market. From this we can deduce that the current world is changing rapidly thanks to technology which is changing our lifestyle in a positive way.

REFERENCES

Christopher, M., & Towill, D. (2001). An integrated model for the design of agile supply chains. *International Journal of Physical Distribution & Logistics Management, 31*(4), 235–246. DOI: 10.1108/09600030110394914

Geunes, J., & Pardalos, P. M. (2003). Network optimization in supply chain management and financial engineering: An annotated bibliography. *Networks, 42*(2), 66–84. DOI: 10.1002/net.10082

Golinelli, G., & Dezi, L. (1997). *Reti, finanza, progetti.* Cedam.

Kagermann, H., Lukas, W. D., & Wahlster, W. (2011). Industrie 4.0: Mit dem Internet der Dinge auf dem Weg zur 4. industriellen Revolution. *VDI nachrichten, 13*(1), 2-3.

La Londe, B. J., & Masters, J. M. (1994). Emerging logistics strategies: Blueprints for the next century. *International Journal of Physical Distribution & Logistics Management, 24*(7), 35–47. DOI: 10.1108/09600039410070975

Nahavandi, S. (2019). Industry 5.0—A human-centric solution. *Sustainability (Basel), 11*(16), 4371. DOI: 10.3390/su11164371

Nonaka, L., Takeuchi, H., & Umemoto, K. (1996). A theory of organizational knowledge creation. *International Journal of Technology Management, 11*(7-8), 833–845.

Ottolia, A. (2017). *Big Data e innovazione computazionale* (Vol. 28). G Giappichelli Editore.

Rullani, E. (1999). Crescita e successione: La metamorfosi del capitalismo personale. *CUOA Rivista, 1*, 13–20.

Teece, D. J. (2000). Strategies for managing knowledge assets: the role of firm structure and industrial context. *Long range planning, 33*(1), 35-54.Itami, H., & Roehl, T. W. (1991). *Mobilizing invisible assets.* Harvard University Press.

Chapter 6
The Innovation of Media Franchises:
Business Models and Industry 5.0 Technologies of Transmedia Content

ABSTRACT

Franchise-based businesses are highly prevalent in the entertainment sector, where intellectual properties (IPs) are continuously expanded across multiple media platforms. This chapter builds upon previous discussions of innovation and knowledge management to explore the evolution of franchising, with a particular focus on international franchising and its various models. Due to the long-running and prolific trend of adapting and managing IPs across different formats and markets, a key emphasis is placed on media franchising. Additionally, the chapter delves into knowledge management within transmedia content, highlighting the role of strategic partnerships in fostering growth through knowledge sharing and commercialization. The impact of emerging technologies on media franchises is explored, particularly in relation to digital transformation, audience engagement, and customer-centric marketing strategies. The chapter concludes with a case study on The Walt Disney Company, analyzing the risks associated with franchise mismanagement.

UNDERSTANDING MEDIA-FRANCHISES: FOUNDATIONS AND KEY CONCEPTS

Internationalization strategies encompass the comprehensive set of decisions and actions a company undertakes to expand its operations beyond domestic borders (Rachmad, 2022). These strategies are fundamental for enterprises aiming to

DOI: 10.4018/979-8-3373-3466-0.ch006

leverage global market opportunities, enhance their competitive edge, and secure sustainable long-term growth. Influential work in this field, notably by Michael Porter in 1985, highlighted that internationalization is far more than a mere geographic expansion—it represents a profound reorganization of the value chain (Tang et al., 2022). This reorganization involves both primary activities (such as logistics, production, marketing, distribution, and after-sales services) and support functions (including human resources, technological development, procurement, and corporate infrastructure).

A central challenge in this process is determining the optimal configuration and coordination of global activities. Companies must decide whether to adopt a decentralized approach—distributing value-chain activities across multiple locations to better adapt to local market demands, despite the risk of duplicated efforts—or a concentrated model that centralizes operations to maximize economies of scale, even if it sacrifices some local flexibility. Various strategic pathways emerge from these choices. For instance, an export-based strategy allows a firm to sell its products in foreign markets while keeping production centralized at home, whereas a decentralized model enables the tailoring of operations to regional specifics. Alternatively, a homogeneous global strategy seeks to exploit interdependencies across countries by standardizing processes and consolidating key functions in the most cost-effective locations, thereby reducing costs and reinforcing a consistent value proposition on a worldwide scale.

The evolving dynamics of international trade have further transformed this landscape. Traditionally focused on the exchange of finished goods, global commerce now frequently involves the segmented trade of production processes. Countries increasingly specialize in distinct phases of the value chain according to their unique resources, expertise, and cost structures. In response, companies often relocate specific production stages to regions where they can achieve greater efficiency, employing long-term supply agreements or joint ventures to share resources, expertise, risks, and returns (Gereffi et al., 2021).

Intercompany agreements have emerged as a pivotal mechanism in this context. By forming strategic alliances—ranging from non-equity contractual collaborations to equity-based ventures such as joint ventures or licensing arrangements—firms can concentrate critical value-adding activities, gain access to specialized know-how, and mitigate risks by sharing the financial burden of market entry. Such cooperative frameworks provide a flexible platform for navigating the complexities of operating across diverse regulatory and cultural environments (Lescrauwaet et al., 2022).

Within this broader framework of international expansion, the franchising system stands out as a distinctive and effective mode of growth. Under a franchising agreement, a franchisor grants a franchisee the right to operate a business using its established brand and management model, in return for an initial fee and ongoing royalties. This arrangement not only transfers valuable organizational know-how and technical expertise to the franchisee but also enables the franchisor to extend its market presence without assuming direct responsibility for every individual outlet. The franchising model finds application across a myriad of sectors—from hospitality and retail to automotive services and financial services.

It is important to distinguish between national and international franchising within this paradigm. National franchising typically involves the expansion of a brand within the familiar context of a single country, where cultural, linguistic, and regulatory environments are consistent, thus simplifying management and reducing risk. In contrast, international franchising requires navigating a mosaic of divergent market dynamics and legal frameworks, which may necessitate significant contractual adaptations and incur higher initial investments due to the uncertainties inherent in entering new territories (Alon et al., 2021).

Historical Overview

Historically, franchising has roots that extend back to ancient times—the Romans, for instance, employed arrangements similar to franchising for managing taverns. However, the modern franchising system began taking shape in the 19th century and gained prominence throughout the 20th century, evolving through a series of pioneering business initiatives.

One early example is the Singer Sewing Machine Company. Founded in 1851 by Isaac Merritt Singer, this American enterprise revolutionized the textile industry by introducing sewing machines that significantly simplified and accelerated both domestic and industrial sewing processes. Singer's early use of licensing agreements for the establishment of retail outlets foreshadowed many of the advantages that modern franchising would later offer, such as increased production efficiency and rapid, large-scale distribution. Over time, the Singer brand became synonymous with quality and reliability, a reputation built in part on its innovative approach to international distribution (Maines, 2023).

Another seminal case is the Coca-Cola Company. In 1899, Coca-Cola launched its franchising program by awarding its first bottling franchise to entrepreneurs in Chattanooga, Tennessee. This initial move set off a swift expansion of an independent bottling network, enabling Coca-Cola to efficiently enter new markets and deeply embed its brand in local communities. The franchising model not only spurred rapid market growth but also forged enduring connections with consumers—an impact that continues to be felt today, as evidenced by the extensive and complex global network of independent bottlers that supports the brand (Felstead, 2024).

Similarly, A&W Root Beer emerged as a pioneering figure in the beverage sector. Its journey began in 1919 with a modest beverage stand in Lodi, California, and took a transformative turn in 1922 when its founder, Roy Allen, converted his establishment into an innovative drive-in restaurant. This move marked the birth of A&W's franchising approach, which allowed the brand to expand swiftly across new markets. The distinctive taste and superior quality of A&W Root Beer soon made it an iconic product, reinforcing the value of a franchising model in building a strong, recognizable brand. (Rice et al., 2023).

Throughout the 20th century, franchising evolved from a relatively novel idea into one of the most widely adopted business models across numerous sectors. Factors such as the increasing demand for standardized products and services, the benefits of economies of scale, and rapid advancements in communication and transportation contributed significantly to this evolution. Companies found franchising to be an effective strategy for expanding their market presence while mitigating the financial and operational risks associated with opening new outlets. As the model matured, franchisors began to centralize control over operations to ensure uniformity in quality and customer experience, even as they extended their reach both domestically and internationally.

The transformation of franchising reached a new pinnacle with the rise of McDonald's. Originally founded in 1940 by Richard and Maurice McDonald, the chain was revolutionized in the 1950s by Ray Kroc, who recognized the enormous potential for national—and eventually international—expansion via franchising. McDonald's emphasis on operational standardization, rigorous quality control, and profit-sharing between franchisor and franchisee not only enabled rapid growth throughout the United States but also established a blueprint for global expansion. This model, along with similar approaches adopted by companies like Hilton Hotels in the hospitality sector, underscored the critical importance of maintaining consistent service and product quality across geographically diverse locations (Alexander, 2023).

Regulatory frameworks evolved alongside these developments. In the 1960s and 1970s, increasing awareness of the need for transparency and franchisee protection led to significant legislative initiatives in the United States. A pivotal moment came in 1979 with the enactment of the Federal Trade Commission Franchise Rule, which required franchisors to provide detailed disclosure documents outlining essential financial, contractual, and operational information. This regulatory milestone not only helped safeguard franchisees from potential fraud but also established a clearer, more accountable framework for franchising practices—a foundation that has been further supported by state-level regulations and the guidance of industry organizations such as the International Franchise Association.

In recent decades, the franchising model has experienced a wave of globalization. Advances in technology and communication, coupled with a more interconnected global economy, have made it easier for established franchisors to expand into international markets. The harmonization of commercial regulations across different jurisdictions has further reduced the complexity and cost of adapting franchising models to new territories. Companies with strong domestic track records have successfully leveraged their experience to enter international markets, tailoring their brands and business models to align with local consumer preferences (Erceg et al., 2021).

Moreover, franchising has diversified far beyond its traditional roots in food and hospitality. Today, it spans a wide array of industries—including financial services, fitness, beauty, and more—demonstrating the model's remarkable adaptability and flexibility. Consumer demand for convenience and a variety of products and services continues to drive this expansion, providing companies with an effective strategy for growth without the need to directly manage every new outlet.

International Franchising

International franchising is a strategic approach that enables a franchisor to expand its business operations into foreign markets by adapting its model to meet the diverse needs and regulatory requirements of each destination. This model not only facilitates the growth of successful commercial ventures by connecting the franchisor with partners who provide the necessary capital and local workforce but also offers several notable advantages. Leveraging local management helps address challenges related to language and cultural differences, ensuring that operations are tailored to regional market conditions. The appointment of a local expert to oversee operations further mitigates the difficulties associated with remote management. Engaging local legal advisors ensures compliance with the destination country's laws, thereby limiting the franchisor's direct liability. In some markets, franchising serves to bypass restrictions or controls imposed on foreign direct investments. This approach also reduces exposure to political uncertainties, such as the risk of expropriation of foreign assets, while simultaneously minimizing the impact of customs duties that often hinder international trade. The methods for implementing international franchising vary according to the organization of the commercial network and specific market conditions, with three primary modalities commonly adopted: master franchising, area development agreements, and direct franchising (Jell-Ojobor et al., 2022).

Master Franchising

In a master franchising arrangement, the franchisor grants a master franchisee the exclusive right to develop and manage the franchise system within a defined territory and time frame (Tucker and Bartkiw, 2022). The master franchisee is empowered to recruit and support sub-franchisees, thereby assuming many of the operational challenges associated with market entry. This model is generally chosen under conditions such as:

- Limited financial resources or expertise on the part of the franchisor.
- High administrative costs and complexity in establishing a standalone operation in the target market.
- The need for a minimal capital investment from the franchisor, with a greater reliance on the master franchisee's financial, human, and experiential resources.

The advantages of master franchising include the transfer of local market challenges—ranging from legal regulations and cultural nuances to language barriers—to the master franchisee. The master franchisee receives exclusive rights to develop the market and, in turn, can establish sub-franchising arrangements. However, this model also presents challenges:

- Selecting the right master franchisee is critical, as the success of the system depends on their business acumen and ability to build a profitable, compliant network.
- The franchisor must set and monitor minimum growth criteria; failure to achieve these targets may result in contractual adjustments such as the suspension of exclusive growth rights or reductions in royalty percentages.
- There is an inherent trade-off in control, as the franchisor must cede significant operational autonomy to the master franchisee, who then assumes direct responsibility for training, distribution, and promotional activities within the territory.

Area Development Agreements (ADAs)

Area development agreements grant a franchisee—acting as the regional developer—the exclusive right to open and manage a predetermined number of franchise units within a specific geographic area and time period (Hadjielias et al., 2021). These agreements are especially useful in large domestic markets where the territory can be divided into several exclusive zones. Key features of ADAs include:

- **Territorial Exclusivity:** The franchisor precisely delineates the area in which the franchisee has exclusive rights, ensuring a focused and competitive development strategy.
- **Multiple Contractual Layers:** Initially, a master contract is established between the franchisor and the area developer. The developer then enters into individual agreements with each franchise unit within the designated area. This dual-layer arrangement allows the franchisor to maintain the option to terminate regional development agreements for underperforming units without disrupting the entire network.
- **Flexibility in Assignment:** In markets where a single large franchisee might not be viable, the franchisor may opt to assign smaller, exclusive territories to several developers. This approach encourages healthy competition and mitigates the risks associated with a single point of failure.

While ADAs can reduce the administrative burden on the franchisor by transferring many operational responsibilities to the area developer, they also require careful oversight. The franchisor must retain control over key contractual stipulations—such as prohibiting unauthorized transfers of territory—and ensure that the developer has the necessary resources, training, and operational support to meet established performance targets.

Direct Franchising

Direct franchising involves the franchisor entering into separate agreements with individual franchisees while providing ongoing, fundamental support throughout the duration of the contract (Bretas and Alon, 2021). This model ensures a higher level of control over each franchise unit, as the franchisor maintains direct relationships with all affiliates. The primary advantages of direct franchising include:

- **Enhanced Control:** With separate contracts for each unit—even when a single franchisee manages multiple locations—the franchisor can closely monitor compliance with operational standards and maintain consistent brand positioning.
- **Market Testing and Adaptation:** Direct franchising serves as an effective pilot project, allowing the franchisor to test and adapt the business model in a new market before considering more complex arrangements such as ADAs.
- **Localized Problem Solving:** This model helps address local challenges, such as linguistic differences, legal and cultural variations, and diverse consumer preferences. The franchisor is better positioned to provide tailored support and resolve issues quickly.

- **Streamlined Distribution and Advertising:** Direct relationships enable the franchisor to implement centralized training programs, establish efficient distribution networks, and coordinate advertising efforts more effectively.

However, direct franchising also comes with challenges. The franchisor must ensure that each potential franchisee—often a small entrepreneur with limited financial and commercial expertise—operates in strict compliance with local laws and established operational standards. Moreover, significant resources must be allocated to initial training, ongoing support, and the establishment of an effective supply chain. Despite these challenges, direct franchising remains a viable and often advantageous model, particularly when the target market is in close proximity to the franchisor's headquarters or when regulatory or fiscal barriers limit more indirect forms of market entry.

Media-Franchising to Sustain Technological Change

A media franchise is a series of interconnected works that span various media formats and entertainment channels. Understanding its key components is essential to grasp how these franchises captivate audiences and generate extensive market opportunities over time (Iyer, 2024). While international franchising centers on replicating and adapting proven business models to various global markets, media franchising takes a different path by leveraging creative content and narrative universes. Instead of expanding through physical outlets and local operational expertise, media franchises build their strength on a core intellectual property that resonates with audiences across multiple platforms. This approach transforms a singular creative work into an expansive ecosystem—one that spans films, television, books, games, and digital content. The following discussion explores the key components that underpin a successful media franchise, highlighting how these elements work together to captivate audiences and unlock diverse market opportunities over time.

- **Core Intellectual Property:** At the heart of every media franchise lies a robust piece of intellectual property—the original work that establishes the universe, characters, and narrative tone. This foundational element, whether it begins as a book, film, or even an online persona, is crucial for drawing in audiences and setting the stage for future expansion. For example, the "Harry Potter" series, which began with J.K. Rowling's debut novel in 1997, not only captured the imagination of millions but also laid the groundwork for films, games, theme parks, and merchandise. Similarly, online personalities like MrBeast have leveraged their initial content success to build diverse business ventures, including virtual restaurant chains and branded merchandise.

- **Emotional Engagement**: The ability to forge a deep emotional connection with the audience is paramount for a media franchise. When viewers or readers feel a personal bond with the characters and stories, they are more likely to remain loyal, invest in related products, and promote the brand organically. This emotional engagement transforms casual consumers into dedicated fans—such as the "potterheads" of the Harry Potter series or the passionate communities supporting franchises like Star Wars and the Marvel Cinematic Universe. In the case of modern digital creators, emotional engagement through innovative content and interactive experiences can directly drive merchandise sales and brand expansion.
- **Distribution Channels and Transmedia Storytelling**: Effective distribution and narrative expansion across multiple platforms are vital for extending a media franchise's reach. Transmedia storytelling uses various channels—films, TV shows, books, comics, video games, and online platforms—to reveal different facets of the overarching narrative. This approach not only enriches the audience's experience but also encourages engagement across a diverse range of media. Platforms such as streaming services, social networks, and dedicated websites like Pottermore help maintain continuous customer engagement and expand the franchise's audience globally.
- **Teaser and Perpetual Marketing Strategies**: Strategic marketing plays a critical role in both launching and sustaining a media franchise. Teaser marketing builds anticipation by releasing cryptic, intriguing snippets of information that spark curiosity and conversation, as famously practiced by companies like Apple during new product launches. In contrast, perpetual marketing focuses on maintaining continuous engagement with the audience through a steady flow of content, personalized communication, and consistent updates. This dual approach not only captures the public's attention at launch but also fosters long-term brand loyalty and repeated interaction.
- **Brand Consistency and Control**: A defining strength of a media franchise is its ability to maintain a coherent brand identity across various platforms and extensions. This involves ensuring that every piece of content, product, or adaptation adheres to established narrative, visual, and thematic standards. Consistency reinforces brand trust and provides fans with a reliable and immersive experience, as seen with franchises like the Marvel Cinematic Universe, Harry Potter, and Pokémon. Active brand control through careful licensing and quality oversight is essential to preserve the integrity of the original concept, even as the franchise expands into new markets and media formats.

THE ROLE OF KNOWLEDGE IN EFFECTIVE MEDIA-FRANCHISE MANAGEMENT

In the contemporary media landscape, the ability to manage a successful franchise depends not only on financial investment and marketing strategies but also on the strategic acquisition and application of knowledge. In an increasingly interconnected and competitive market, where consumer expectations evolve rapidly, a deep understanding of brand positioning, audience engagement, and cross-platform integration is essential. By leveraging knowledge effectively, media franchises can expand their reach, strengthen their brand identity, and optimize resource utilization to create content that stands out in a saturated market.

Knowledge management in media franchises encompasses various domains, from intellectual property rights and storytelling continuity to audience analytics and technological advancements. Successful franchises integrate these elements strategically, ensuring that every aspect of brand expansion aligns with the overall vision and market expectations. Whether developing new content, engaging in licensing agreements, or establishing cross-industry partnerships, franchises must navigate a complex ecosystem that requires informed decision-making and adaptability (Gorovaia et al., 2023).

A well-managed franchise capitalizes on strategic partnerships to enhance its market presence and consumer appeal. By collaborating with complementary brands, franchises can extend their influence, reach new demographics, and offer unique value propositions that would be difficult to achieve independently. Such collaborations manifest in various forms, including co-branded products, shared advertising campaigns, and interactive events, each contributing to the franchise's visibility and consumer engagement.

The effectiveness of franchise partnerships largely depends on the alignment of goals and values between stakeholders. A shared vision ensures that all parties work cohesively toward common objectives, avoiding conflicts that may dilute the brand's integrity. These collaborations can vary in scope and duration, from short-term promotional alliances to long-term strategic relationships designed to sustain brand relevance over time. A crucial aspect of these partnerships is the establishment of clear roles, expectations, and responsibilities to prevent misunderstandings and ensure that the collaboration generates value that neither entity could achieve independently (Raha and Hajdini, 2023).

Throughout the years, several case studies have demonstrated the power of strategic alliances within the media franchise industry. A notable example is the collaboration between entertainment giants Marvel and Fortnite, which seamlessly integrated popular superheroes into the gaming world, creating a crossover experience that captivated audiences across both fan bases. Another successful case is the partnership between LEGO and various film franchises, which has led to interactive products that engage consumers on multiple levels—through toys, digital experiences, and cinematic adaptations. These collaborations exemplify how knowledge-driven partnerships can amplify brand impact, generate new revenue streams, and enhance audience engagement.

Strategic alliances within media franchises can take on multiple forms, each offering distinct advantages. Co-branding enables franchises to merge their strengths, resulting in innovative products that combine established identities in a fresh, compelling way. Joint advertising campaigns optimize marketing efforts, broadening reach while minimizing individual promotional costs. Event collaborations, such as themed pop-up experiences or co-hosted conventions, create immersive interactions that foster deep audience connections and long-term loyalty. Licensing agreements allow franchises to expand into new markets while maintaining brand consistency, while cross-promotional efforts leverage each partner's platform to mutually boost visibility and engagement. The selection of the most suitable collaboration strategy depends on the franchise's objectives, target audience, and long-term vision (Gaitán, 2021).

One of the key benefits of knowledge-driven franchise management is its ability to maximize audience engagement and market penetration. By partnering with brands that share similar target demographics, a franchise can tap into established consumer bases, enhancing credibility and broadening its appeal. Moreover, strategic alliances facilitate resource-sharing, allowing franchises to optimize production capabilities, distribution channels, and marketing expertise while reducing operational costs. The integration of complementary technologies, storytelling techniques, and market insights strengthens the overall brand proposition, making it more resilient in a competitive environment.

To maximize the advantages of collaborative franchise management, selecting the right partners is essential. The most effective alliances are built on a foundation of shared values, aligned objectives, and complementary brand attributes. Compatibility between the respective consumer bases is another crucial factor, as it ensures that the collaboration resonates with the target audience. Additionally, well-defined partnership goals contribute to a structured and transparent collaboration, minimizing conflicts and ensuring a cohesive brand message.

Despite its numerous advantages, franchise collaboration presents challenges that must be addressed strategically. Differences in creative direction, operational methodologies, and long-term priorities can lead to friction between partners. Effective communication, clear contractual agreements, and structured conflict resolution mechanisms are essential to mitigating these risks. Maintaining brand integrity is another critical concern, as misaligned partnerships can dilute a franchise's identity and weaken consumer trust. Measuring the success of franchise collaborations also requires well-defined key performance indicators (KPIs), including audience engagement metrics, revenue growth, and brand sentiment analysis (Helmold, 2022).

Ultimately, the effective management of media franchises relies on a well-rounded approach that integrates strategic knowledge application, cross-industry partnerships, and continuous market adaptation. By leveraging intellectual property insights, audience behavior analytics, and technological advancements, franchises can sustain their relevance and drive long-term success. As emerging trends such as digital transformation, sustainability-conscious branding, and influencer-driven marketing continue to shape the industry, franchises that invest in knowledge-driven strategies will be best positioned to navigate future challenges and seize new opportunities.

The evolution of franchise partnerships highlights the importance of continuous learning and strategic adaptability. As media landscapes grow increasingly complex, franchises that prioritize knowledge management will not only strengthen their brand position but also create innovative, engaging experiences that resonate across generations. By embracing a knowledge-centric approach to franchise management, businesses can maximize their competitive edge and establish enduring consumer connections in an ever-evolving industry landscape.

LEVERAGING NEW TECHNOLOGIES IN TRANSMEDIA ENTERTAINMENT

Mass communication has played a pivotal role in shaping modern society by influencing cultural values, economic trends, and political dynamics. Over centuries, media have evolved through transformative stages—from the advent of the printing press, which revolutionized information dissemination, to the emergence of newspapers that provided timely updates on societal affairs. The subsequent rise of radio and television further expanded access to real-time news and entertainment, fostering a shared public consciousness while stimulating economic growth through burgeoning industries in publishing and advertising (Priadi and Thariq, 2023).

In recent decades, the digital revolution has dramatically reconfigured the media landscape. The proliferation of the Internet and digital platforms has enabled instantaneous, global information exchange, fundamentally altering consumer engagement. Social media, on-demand streaming, and advanced content creation tools have collectively paved the way for transmedia entertainment—a model where narratives unfold seamlessly across multiple platforms. This convergence of technologies empowers creators to craft immersive, interconnected stories that resonate with a digitally sophisticated audience. By harnessing these innovations, stakeholders not only extend market reach but also reinvent storytelling methods, creating a dynamic and multifaceted entertainment ecosystem that continuously adapts to evolving audience preferences (Unnava and Aravindakshan, 2021).

The foundation of mass communication can be traced back to the development of the printing press, which marked a profound shift in the dissemination of knowledge. The invention of movable-type printing by Johannes Gutenberg in the 15th century represented a technological breakthrough that facilitated the mass production of books and newspapers. Prior to this innovation, manuscripts were manually copied, limiting their availability to a select few. By reducing the cost and time required for reproducing texts, the printing press enabled a broader segment of society to access literature, contributing significantly to rising literacy rates. This advancement played a crucial role in major cultural and intellectual movements such as the Renaissance and the Protestant Reformation, where the circulation of written ideas spurred profound societal change. Moreover, the emergence of the printing industry stimulated economic growth, creating employment opportunities in publishing, bookbinding, and the production of paper and ink. The ability to distribute written content on a large scale also contributed to the early frameworks of globalization, as economic and scientific knowledge could be shared across regions and continents (Nyambane, 2021).

As mass communication evolved, newspapers emerged as a dominant medium during the 18th and 19th centuries. The invention of rotary printing, patented by Richard March Hoe in 1843, revolutionized newspaper production by enabling the rapid printing of thousands of copies within hours. This technological advancement made newspapers widely accessible and affordable, fostering the rise of modern journalism. With the ability to provide regular updates on political, economic, and social affairs, newspapers became instrumental in shaping public opinion and disseminating democratic ideals. Additionally, the newspaper industry contributed to economic development by generating revenue through sales and advertisements. Businesses utilized print media to reach broad audiences, promoting consumerism and facilitating market expansion. By publishing information on commodity prices, employment opportunities, and investment prospects, newspapers also played a key role in enhancing economic efficiency and commerce.

The 20th century witnessed a paradigm shift in mass communication with the introduction of electronic media. The advent of radio broadcasting in the early 1920s allowed for the real-time transmission of news and entertainment to a global audience (Medoff and Kaye, 2021). As radio rapidly gained popularity, it became a primary source of information and cultural expression, particularly during significant historical events such as World War II, where it served as a crucial tool for public information and propaganda.

Television, which emerged as a dominant medium in the 1950s, further transformed mass communication by combining visual and auditory elements. This medium not only revolutionized entertainment but also had a profound impact on public discourse, as televised news coverage brought political and social issues into the homes of millions. The expansion of television networks contributed to economic growth by creating a robust industry encompassing broadcasting, advertising, and content production. Furthermore, television's influence extended to related sectors such as film and music, enabling these industries to reach broader audiences through televised programming. Advertising via television also played a significant role in driving consumer behavior, fueling economic expansion by promoting products and services on a national and international scale (Arnold, 2021).

In recent decades, the digital revolution has dramatically reconfigured the media landscape. The proliferation of the Internet and digital platforms has enabled instantaneous, global information exchange, fundamentally altering consumer engagement. Social media, on-demand streaming, and advanced content creation tools have collectively paved the way for transmedia entertainment—a model where narratives unfold seamlessly across multiple platforms. This convergence of technologies empowers creators to craft immersive, interconnected stories that resonate with a digitally sophisticated audience. By harnessing these innovations, stakeholders not only extend market reach but also reinvent storytelling methods, creating a dynamic and multifaceted entertainment ecosystem that continuously adapts to evolving audience preferences (Lazzeretti, 2023).

The integration of new digital tools has also expanded economic opportunities within the media industry. Digital advertising has become a dominant revenue stream, with businesses leveraging data analytics and targeted marketing to reach specific audiences. Additionally, content monetization through subscription services, digital publications, and influencer-driven branding has diversified revenue sources. As media continue to evolve, their role in shaping public perception, fostering economic growth, and influencing cultural development remains more significant than ever (Williams, 2021).

The advent of digital technology, initiated with the development of the Internet during the 1960s and 1970s and significantly accelerated in the 1990s with the rise of the World Wide Web, has fundamentally reshaped the processes of information creation, distribution, and consumption. Initially conceived as a military research initiative in the United States, the Internet evolved into a global communication network that has radically transformed interpersonal interactions, professional environments, and access to knowledge. The introduction of the World Wide Web in 1989 marked a pivotal moment, facilitating widespread public access to online resources and laying the foundation for a digitally interconnected world (Akhtar et al., 2024).

The 2000s witnessed a further shift in the media landscape with the emergence of social networking platforms such as Facebook, Twitter, Instagram, and YouTube. These digital spaces democratized content creation, enabling individuals to disseminate information and entertainment on an unprecedented scale. This transformation fostered the rise of digital influencers and the phenomenon of viral content, significantly impacting marketing strategies and consumer engagement. The accessibility of content production has had multifaceted implications. On the one hand, social media personalities play a crucial role in shaping public discourse and cultural trends, often receiving early access to products to generate anticipation and discussion among their followers. Additionally, their influence extends to sustaining audience interest in intellectual properties by speculating on future developments within established franchises. On the other hand, the ease of content dissemination poses challenges regarding intellectual property rights. The unauthorized disclosure of proprietary information, often facilitated by individuals known as "leakers," represents a significant issue. These individuals release confidential data pertaining to corporations, governments, or entertainment industries, ranging from internal documents and strategic plans to unannounced product details and state secrets. Such leaks can exert profound effects on public perception, corporate reputations, and political landscapes, sometimes leading to significant economic or diplomatic consequences. The low-cost nature of digital content production has also contributed to the proliferation of competing works, including those that incorporate or derive inspiration from pre-existing intellectual properties without authorization (Burns, 2021).

The transformation brought about by digitalization extends beyond content creation to the modes of consumption. The rise of streaming services and on-demand platforms has revolutionized the entertainment industry, allowing audiences to access a vast array of audiovisual content at their convenience. Companies such as Netflix, Amazon Prime, and Spotify have redefined the distribution model by placing control directly in the hands of users, enabling personalized viewing and listening experiences unbound by traditional broadcasting schedules.

The economic impact of the digital revolution has been profound, particularly within the technology sector. Companies like Google, Apple, Microsoft, and Amazon have ascended to dominant positions within the global economy, generating billions in revenue and creating employment opportunities across multiple industries. The proliferation of technology startups has been fueled by access to venture capital and digital infrastructure, driving innovation in commerce, communication, healthcare, and education. E-commerce exemplifies the transformative power of digitalization. Online marketplaces such as Amazon, eBay, and Alibaba have revolutionized retail by breaking down geographical barriers and enabling small enterprises to engage with international markets. While this shift has challenged traditional brick-and-mortar businesses, necessitating adaptation or closure for many, it has simultaneously fostered efficiency, expanded consumer choices, and introduced cost-effective business models that continue to reshape global commerce (Vassolo et al., 2024).

The diversification of communication channels represents a key element in contemporary marketing strategies, particularly in the franchise sector. Technological evolution has expanded the media landscape, providing companies with increasingly sophisticated tools to engage with consumers. Industry 5.0, with its emphasis on synergy between artificial intelligence, advanced automation, and human interaction, has revolutionized communication strategies, enabling more effective personalization and seamless integration across various marketing channels.

The adoption of intelligent technologies has transformed how franchises promote their products and services. Artificial intelligence and machine learning allow for in-depth analysis of consumer behavior, making it possible to tailor advertising messages to individual preferences. Big data systems collect and analyze information in real time, ensuring greater accuracy in audience segmentation and marketing campaign personalization.

Another innovative aspect introduced by Industry 5.0 is the use of immersive technologies such as augmented reality (AR) and virtual reality (VR). These tools offer new engagement methods, allowing consumers to interact with brands in personalized virtual environments. For example, companies can create interactive experiences where customers virtually explore new products or participate in exclusive events in digital spaces. These technologies are particularly effective in experiential marketing strategies, strengthening the emotional bond between the brand and the consumer.

The convergence of digital platforms and automation has also fostered the development of omnichannel marketing. Industry 5.0 enables seamless integration of social media, email marketing, programmatic advertising, virtual assistants, and intelligent chatbots, providing a cohesive experience across all touchpoints. In the franchise sector, this strategy translates into more effective and adaptable communication, ensuring that every consumer interaction is relevant and timely (Grech et al., 2023).

Another advantage brought by Industry 5.0 is the use of cobots (collaborative robots) to optimize the production and distribution of marketing content. Cobots can assist in the automated generation of advertising materials, the analysis of campaign performance, and the management of customer interactions, thereby improving the overall efficiency of communication strategies.

The adoption of these technologies is reshaping the marketing sector in franchises, offering new opportunities for differentiation and audience engagement. The ability to combine advanced digital tools with a user-centered approach allows companies to create tailored experiences, increasing customer loyalty and brand awareness. Industry 5.0 not only enhances communication effectiveness but also redefines the role of the consumer, transforming them into an active participant within the brand ecosystem.

Technological Innovation and Consumer-Centric Marketing

In the digital age, the diversification of communication channels has become a defining feature of modern marketing strategies. As businesses adapt to an increasingly interconnected world, the ways in which brands engage with consumers have undergone significant transformation. The advent of the internet and the proliferation of digital platforms have facilitated a shift from traditional, one-directional advertising to interactive and highly personalized communication strategies.

One of the most notable changes in recent years has been the transition from mass media-driven marketing to multi-channel and omnichannel approaches. Initially, brands primarily relied on television, radio, and print advertising to reach broad audiences. However, the rise of digital platforms, particularly social media, has enabled companies to engage consumers through tailored content and direct interactions. This evolution reflects a growing emphasis on consumer-centric marketing, where businesses aim to establish meaningful relationships rather than merely promoting products.

The fragmentation of audiences across multiple digital environments has necessitated the adoption of diversified marketing strategies. Companies now leverage various platforms—social media, search engines, email marketing, mobile applications, and even virtual and augmented reality experiences—to maintain visibility and relevance. This multi-channel approach ensures that brands can reach consumers at different touchpoints throughout their decision-making journey, fostering engagement and brand loyalty.

Social media, in particular, has played a pivotal role in reshaping consumer-brand interactions. Unlike traditional advertising, which primarily functions as a one-way communication tool, social media platforms facilitate two-way engagement. Consumers are no longer passive recipients of marketing messages; instead, they actively participate in brand conversations, provide feedback, and even co-create content. This participatory culture has given rise to influencer marketing, user-generated content campaigns, and interactive brand storytelling, all of which enhance authenticity and trust (Sheoran and Vij, 2023).

Moreover, the integration of big data and artificial intelligence (AI) has further refined marketing strategies. Advanced analytics allow businesses to collect, process, and interpret vast amounts of consumer data, enabling hyper-personalized marketing initiatives. AI-powered algorithms analyze browsing behaviors, purchase histories, and social interactions to deliver customized recommendations and targeted advertisements. As a result, brands can anticipate consumer needs more effectively, providing a seamless and personalized experience that aligns with individual preferences.

Another critical development in modern marketing is the increasing reliance on immersive technologies, such as augmented reality (AR) and virtual reality (VR). These innovations offer consumers interactive and experiential brand engagements, bridging the gap between digital and physical interactions. For instance, AR applications enable customers to visualize products in real-world settings before making a purchase, enhancing confidence and reducing uncertainty. Similarly, VR experiences provide immersive storytelling opportunities, allowing brands to craft compelling narratives that resonate deeply with their audiences (Gopakumar and Dananjayan, 2023).

The growing importance of mobile-first strategies further underscores the necessity of adapting to evolving consumer behaviors. With the widespread use of smartphones, businesses must optimize content for mobile consumption, ensuring seamless navigation and engagement across devices. Mobile apps, in particular, serve as direct communication channels, offering personalized promotions, push notifications, and location-based marketing to enhance user experience and retention.

As the marketing landscape continues to evolve, ethical considerations have become increasingly relevant. The rise of data-driven marketing raises concerns regarding consumer privacy and data protection. Regulations such as the General Data Protection Regulation (GDPR) and the California Consumer Privacy Act (CCPA) have compelled businesses to adopt transparent data collection practices, emphasizing the importance of consumer trust in the digital ecosystem. Companies must balance personalization with privacy, ensuring that their strategies align with ethical guidelines and regulatory requirements.

Additionally, sustainability and corporate social responsibility (CSR) have gained prominence as influential factors in consumer decision-making. Modern consumers increasingly prioritize brands that demonstrate environmental and social consciousness. Consequently, businesses are integrating sustainable marketing practices, such as eco-friendly packaging, ethical sourcing, and socially responsible campaigns, to align with evolving consumer expectations. These initiatives not only enhance brand reputation but also foster deeper emotional connections with audiences (Karwowski and Raulinajtys-Grzybek, 2021).

The evolution of marketing strategies in the digital age significantly impacts franchising by reshaping how brands engage with consumers. Franchisors must leverage multi-channel and omnichannel approaches to ensure consistent brand visibility across diverse platforms, catering to the fragmented digital landscape. The integration of big data and artificial intelligence allows franchise systems to offer personalized, localized marketing, enhancing customer engagement and loyalty. Social media, influencer marketing, and user-generated content provide opportunities for franchisees to foster authentic, community-based interactions with customers. Moreover, immersive technologies like augmented and virtual reality can enhance customer experiences, setting franchises apart in a competitive market. Mobile-first strategies are essential for optimizing customer engagement through smartphones, offering personalized promotions and loyalty programs. Ethical marketing practices, ensuring transparency and consumer privacy, are increasingly important in maintaining brand reputation. Furthermore, sustainability and corporate social responsibility initiatives align with consumer values, enhancing brand appeal and fostering deeper customer connections.

CASE STUDY: THE WALT DISNEY COMPANY AND THE DANGERS OF THE STREAMING BUSINESS

The Walt Disney Company, founded by brothers Walt and Roy Disney on October 16, 1923, has grown from humble beginnings into a global entertainment powerhouse. In the 1920s, Walt, after gaining experience in animation, started the Laugh-O-Gram Studio, which went bankrupt in 1923. He moved to Hollywood, joined by Roy, and secured a contract for a series combining live-action and animation, "Alice Comedies," marking the start of Disney Brothers Studio. The company's first major success came with Oswald the Lucky Rabbit in 1927, but after losing rights to the character in 1928, Walt and Ub Iwerks created Mickey Mouse, a game-changer for the company.

By the 1930s, Disney was refining animation with characters like Minnie Mouse and Donald Duck. The company's breakthrough was the 1937 release of *Snow White and the Seven Dwarfs*, the first full-length animated feature, which set new standards and financed future projects. The following decades saw the release of numerous animated classics, including *Pinocchio*, *Fantasia*, and *Bambi*. During World War II, Disney produced propaganda films but continued making entertainment, facing internal struggles like the 1941 strike.

The 1950s brought significant expansion, including the launch of Disneyland in 1955 and successful films like *Cinderella* and *Peter Pan*. The company's television ventures also flourished with programs like *Disneyland* and *The Mickey Mouse Club*. In the 1960s, Disney expanded its film and theme park operations, including plans for Walt Disney World, which opened posthumously after Walt's death in 1966. The company continued its success with films like *Mary Poppins* and *The Sword in the Stone*.

The 1980s marked a turning point with CEO Michael Eisner's leadership, reviving Disney's fortunes with *The Little Mermaid* (1989). This era also saw the company expand globally with Tokyo Disneyland in 1983. The 1990s ushered in the "Disney

Renaissance," producing hit films like *The Lion King* and *Aladdin*, along with the acquisition of ABC, Pixar, and later Marvel, solidifying Disney's media empire.

In the 2000s, Disney embraced digital animation with *Monsters, Inc.* and acquired Pixar, followed by other strategic acquisitions like Marvel and Lucasfilm. The 2010s were marked by blockbuster films like *Frozen*, the dominance of the Marvel Cinematic Universe, and the launch of Disney+ in 2019. Despite challenges like the COVID-19 pandemic, Disney continues to evolve, focusing on diversity and global content distribution (Mandracchia, 2021).

Disney has effectively utilized various media channels to build its global presence and create a media empire. The company leverages both traditional and modern means of brand promotion to maintain and strengthen its identity worldwide.

In the realm of traditional advertising, Disney has made extensive use of TV commercials to promote its films, especially animated features. A noteworthy example of this is the 2002 campaign for "Lilo & Stitch," where TV spots featured the alien character Stitch interacting with beloved Disney characters from older films. This creative approach not only appealed to long-time Disney fans by introducing Stitch into the established universe but also played into the cultural shift brought on by movies like *Shrek*, which had already begun to deconstruct fairy tale tropes. Radio advertisements, while less prominent than TV ads, were also employed to advertise films and Disney theme parks, reaching audiences who might not be as engaged with visual media. Print advertising has been a staple for promoting Disney's cinematic releases as well. Movie posters for iconic Disney films such as *Snow White and the Seven Dwarfs* and *Cinderella* have transcended their original purposes to become iconic works of art, often found on merchandise like mugs and cushions. Magazines aimed at parents and families frequently feature ads for Disney films and theme parks, while newspapers in the pre-digital age were crucial in publicizing events like the opening of Disneyland and the debut of new attractions. Additionally, billboards in high-traffic areas like Times Square in New York or Sunset Boulevard in Los Angeles have long been used to advertise Disney films and shows, aiming to capture the attention of massive audiences and create brand awareness.

In the realm of digital marketing, Disney has embraced the internet as a central hub for its brand. Disney.com serves as a gateway to all the company's properties, linking users to Disney Parks, Disney Channel, Disney Plus, and more. The website is user-friendly and designed to offer a seamless experience, allowing visitors to easily find information, watch trailers, and purchase products. For major franchises like Marvel, Star Wars, and Pixar, dedicated websites provide exclusive content such as interviews, concept art, and news, keeping fans engaged and updated. Disney has also created an integrated social media strategy across platforms like Facebook, Instagram, Twitter, TikTok, and YouTube. Each platform serves a different function, with Instagram focusing on visual content, Twitter providing real-time updates,

TikTok sharing engaging, viral content, and YouTube showcasing trailers and longer videos. The company actively engages with its fanbase, encouraging them to participate in polls, quizzes, and share user-generated content. A notable example of this is the 2018 animated film *Ralph Breaks the Internet*, which incorporated the now-defunct "Oh My Disney" website into its storyline, inviting viewers to take a quiz to find out which Disney princess they resembled most. Furthermore, Disney uses paid ads on platforms like Google and Facebook to target specific audience segments, promoting new film releases, Disney Plus launches, and discounts for theme parks. Retargeting is also a key strategy, showing personalized ads to users who have previously interacted with Disney content, aiming to drive conversions like ticket sales or merchandise purchases. Disney's content marketing extends to running blogs on various platforms, offering family-friendly educational and entertaining content, as well as updates on Disney Parks and behind-the-scenes stories (Liu J. and Liu Y., 2024).

Public relations (PR) efforts also play a critical role in Disney's brand strategy. For film and attraction launches, Disney issues press releases that are picked up by media outlets worldwide, ensuring extensive coverage. An example of a successful PR campaign was the announcement of Disney Plus, which generated global media buzz. Disney uses PR not only to promote new initiatives but also to manage its public image, responding to crises and controversies through official statements and interviews. High-profile events like the D23 Expo, a bi-annual fan convention, serve as major platforms for announcing new films and projects, strengthening the bond with Disney's fanbase. Additionally, movie premieres are media spectacles, often featuring celebrities on the red carpet, live interviews, and global media coverage. These events significantly amplify the Disney brand's visibility.

When it comes to event marketing, Disney participates in major industry conventions like the San Diego Comic-Con, where it showcases exclusive trailers and footage, hosts panels with filmmakers and stars, and engages directly with fans. Beyond fan conventions, Disney also attends industry events like CinemaCon, where it presents upcoming projects to distributors and industry professionals. In terms of sponsored events, Disney has utilized the Super Bowl to air trailers for blockbuster films like *Star Wars* and *The Avengers*, ensuring global visibility and social media buzz. The company also sponsors cultural events like film festivals and charity initiatives such as Make-A-Wish, blending entertainment with social responsibility and creating emotional connections with audiences.

Merchandising is another core element of Disney's marketing strategy. The company leads the toy industry with products based on its films and characters, selling them worldwide. Collaborations with companies like Hasbro and Mattel have led to successful toy lines, such as *Star Wars* action figures and *Frozen* dolls. Disney has also partnered with fashion brands, creating clothing collections for both children and adults, and even collaborating with luxury brands like Gucci and Coach on Disney-themed high-end fashion lines. Additionally, Disney has licensed its brand to various sectors, allowing for co-branded products like the popular Disney-themed shoes from Vans (Zhu, 2024).

In recent years, influencer marketing has become a significant component of Disney's promotional strategy. The company collaborates with influencers across platforms like YouTube, Instagram, and TikTok to promote its films and theme parks. For instance, travel influencers may be invited to Disneyland to document their experience, reaching millions of followers and driving brand awareness. Celebrity endorsements also play a crucial role in Disney's campaigns, with stars voicing characters in animated films—such as Robin Williams as the Genie in *Aladdin* and Tom Hanks as Woody in *Toy Story*—and serving as brand ambassadors in advertising campaigns. A prime example of this is the "I'm Going to Disneyland!" campaign, which featured celebrities and athletes announcing their visits to Disney parks, linking the brand to celebrations and victorious moments (Joshi et al., 2023).

Disney also utilizes email marketing to keep subscribers informed about new releases, special offers, and exclusive Disney Plus content. Personalized newsletters are sent to subscribers based on their interests, such as specific Disney franchises like *Marvel*, *Pixar*, or *Star Wars*. Disney uses email campaigns to promote special offers during holiday seasons or events like Disney Day, providing incentives like exclusive discounts or early access to new products. Furthermore, these emails are used to gather feedback from customers, which helps Disney improve its offerings while strengthening its relationship with its audience.

Search engine optimization (SEO) and search engine marketing (SEM) are key digital strategies employed by Disney. The company ensures that its websites are optimized for search engines, incorporating relevant keywords and regularly updated content. This ensures that when people search for information on Disney films, characters, or attractions, the company's official websites appear at the top of the results. Disney also invests in SEM, using paid advertising to promote new films, Disney Plus subscriptions, and theme park tickets. Remarketing is another tactic used to re-engage users who have visited the website without completing a purchase, showing them tailored ads to encourage conversion.

Finally, word-of-mouth and buzz marketing are crucial aspects of Disney's strategy. The company aims to create unforgettable experiences in its theme parks and through its films, encouraging people to share their experiences with friends and family. This organic promotion has been particularly effective with films like *Frozen*, which saw a surge of positive word-of-mouth contributing to its global success. Disney also creates buzz through surprise product launches, unexpected trailer releases, and exclusive merchandise, as seen with the surprise launch of Disney Plus in international markets.

In addition, Disney's loyalty programs, such as the Disney Vacation Club and the Disney Rewards Visa Card, offer incentives for customers to visit its parks and engage with its services more frequently. The company also runs promotions and discounts for park tickets and merchandise, often tied to special events or holidays, to encourage repeat business and deepen brand loyalty (Moulton, 2022).

In the 2010s, the rise of streaming services dramatically transformed the entertainment industry and changed the way audiences consume audiovisual content. Platforms like Netflix provided on-demand access to content, allowing viewers to choose what to watch, when to watch it, and on any device. This shift granted viewers increased freedom, as they could now bypass traditional television schedules and select content tailored to their preferences and availability. The phenomenon of binge-watching, where viewers watch multiple episodes in a single session, became widespread, influencing the production and narrative styles of TV shows, which became more serialized and engaging. However, traditional television and cinema have seen a decline in viewers, as an increasing number of individuals have abandoned conventional viewing methods in favor of streaming.

Streaming services also introduced greater diversification, offering global access to international content. These platforms, not only distributors but also producers, have created original series that have captured both audience attention and industry acclaim. This shift has heightened competition, changing the dynamics of how content is created, distributed, and consumed. In response to these changes, Disney launched Disney+ in 2019, aiming to take control of the distribution of its vast portfolio of valuable intellectual properties, including Marvel, Star Wars, Pixar, and its classic Disney productions. This strategic move allowed Disney to offer a comprehensive service that attracted a wide range of audiences, positioning itself as a strong competitor in the streaming market, which was previously dominated by giants like Netflix and Amazon Prime. Acquiring BAMTech provided Disney with the technological infrastructure to manage a global streaming service, while purchasing 21st Century Fox expanded its content library and bolstered its control over Hulu, another important player in the streaming space.

By the end of its first year, Disney+ had surpassed expectations, reaching 86.8 million subscribers. However, by the final quarter of 2023, Disney+ experienced a loss of 1.3 million subscribers due to a significant price hike. The service has also incurred over 11.4 billion dollars in operational losses since its launch, and it is not expected to turn a profit until late 2024. This stagnation can be attributed to multiple factors, including the end of the COVID-19 pandemic, which had provided a significant boost to the streaming sector. The loss of rights to broadcast Indian Premier League cricket matches also caused a significant drop in subscribers from India, a key market for Disney+. The primary challenges, however, remain the devaluation of Disney's brands and the difficulty of maintaining consistent user interest in the platform.

During the pandemic, films were released directly to streaming platforms, bypassing traditional theatrical releases, which allowed Disney to avoid splitting revenues with cinemas and achieve greater profits. However, in the long term, this strategy proved detrimental, particularly for brands that did not even receive limited theatrical distribution. Pixar films, for example, were released during fiscal periods lacking other content, without additional charges to subscribers. This turned these films into mere filler, losing the event-like quality that made them major family spectacles. Post-pandemic box office returns for Pixar films, especially *Lightyear*, the *Toy Story* spin-off, reflected a public disinterest, with the film grossing only 226.4 million dollars on a 200-million-dollar budget (Tian, 2024).

In the case of Disney, its powerful brand recognition works against it. Audiences, already conditioned to expect films like *The Little Mermaid* or *Marvel* superhero productions to eventually appear on Disney+, lose incentive to watch these films in theaters. This issue is less pronounced for other companies like Paramount, which do not have the same level of brand recognition. For example, audiences did not automatically associate films like *Top Gun: Maverick* or *Sonic the Hedgehog 2* with Paramount+ and thus prioritized theater releases over streaming options.

This issue is part of a broader trend in which Disney has accelerated content production since the launch of Disney+. For example, prior to the pandemic, the Marvel Cinematic Universe (MCU) typically released a maximum of two films per year, with 23 films produced between 2008 and 2019. Post-pandemic, however, the MCU has released 11 films and 12 TV series from 2021 to 2024 (Chen et al., 2022). This accelerated production, coupled with the in-house creation of series by Marvel Studios, has led to concerns over oversaturation and declining quality. The same issues apply to Star Wars and Pixar, where new content, particularly on Disney+, has failed to attract sufficient audience engagement or expand the franchises in meaningful ways. These once-dominant franchises are now suffering from *franchise fatigue*, as evidenced by content oversaturation, declining product quality, and a shift in audience tastes, especially with the growing success of video game adaptations in film, which may soon overshadow superhero films as the primary cultural phenomenon in audiovisual media.

REFERENCES

Akhtar, M. N., Haleem, A., & Javaid, M. (2024). Exploring the advent of Medical 4.0: A bibliometric analysis systematic review and technology adoption insights. *Informatics and Health*, *1*(1), 16–28. DOI: 10.1016/j.infoh.2023.10.001

Alexander, T. (2023). Unwrapping the McDonald's model: An introduction to dynamic social theory. *Journal of American Culture*, *46*(3).

Alon, I., Apriliyanti, I. D., & Henríquez Parodi, M. C. (2021). A systematic review of international franchising. *Multinational Business Review*, *29*(1), 43–69. DOI: 10.1108/MBR-01-2020-0019

Arnold, S. (2021). *Gender and Early Television: Mapping Women's Role in Emerging US and British Media, 1850-1950*. Bloomsbury Publishing. DOI: 10.5040/9781350192478

Bretas, V. P. G., & Alon, I. (2021). Franchising in Brazil: Foreign entrants build a presence. *The Journal of Business Strategy*, *42*(6), 374–381. DOI: 10.1108/JBS-03-2020-0051

Burns, K. S. (2021). The history of social media influencers. *Research perspectives on social media influencers and brand communication*, 1-22.

Chen, X., Liu, X., & Wang, X. (2022, November). Comparative Analysis of Disney+ and Netflix in Post-covid19 Era. In *2022 International Conference on Science Education and Art Appreciation (SEAA 2022)* (pp. 1229-1238). Atlantis Press.

Erceg, A., Bilos, A., & Kelic, I. (2021). Franchising in the digital era–What lies ahead. *Economic and Social Development: Book of Proceedings*, 53-65.

Felstead, A. (2024). *The corporate paradox: Power and control in the business franchise*. Taylor & Francis. DOI: 10.4324/9781003527305

Gaitán, M. G. (2021). Licensing as a Central Structure of Technology Transfer Agreements–Joint Venture and Franchising Agreements. *GRUR International*, *70*(5), 427–439. DOI: 10.1093/grurint/ikab032

Gereffi, G., Lim, H. C., & Lee, J. (2021). Trade policies, firm strategies, and adaptive reconfigurations of global value chains. *Journal of International Business Policy*, *4*(4), 506–522. DOI: 10.1057/s42214-021-00102-z

Gopakumar, S., & Dananjayan, M. P. (2023). Augmented reality in modern marketing: The ultimate solution to engage the digitally overwhelmed, budget-conscious audience?. *Journal of Information Technology Teaching Cases*, 20438869231202712.

Gorovaia, N., Pajić, M., & Windsperger, J. (2023). Performance of knowledge transfer mechanisms: The case of franchising networks. *Industrial Marketing Management*, *115*, 539–549. DOI: 10.1016/j.indmarman.2023.09.019

Grech, A., Mehnen, J., & Wodehouse, A. (2023). An extended AI-experience: Industry 5.0 in creative product innovation. *Sensors (Basel)*, *23*(6), 3009. DOI: 10.3390/s23063009 PMID: 36991718

Hadjielias, E., Dada, O. L., & Eliades, K. (2021). Entrepreneurial process in international multiunit franchise outlets: A social capital perspective. *Journal of Business Research*, *134*, 13–28. DOI: 10.1016/j.jbusres.2021.05.022

Helmold, M. (2022). *Performance excellence in marketing, sales and pricing. Berliini*. Springer. DOI: 10.1007/978-3-031-10097-0

Iyer, S. (2024). From Film to Franchise: The Rise of Cinematic Universes and Their Impact on Global Cinema. *Shodh Sagar Journal of Language, Arts. Culture and Film*, *1*(2), 14–20.

Jell-Ojobor, M., Alon, I., & Windsperger, J. (2022). The choice of master international franchising–A modified transaction cost model. *International Business Review*, *31*(2), 101942. DOI: 10.1016/j.ibusrev.2021.101942

Joshi, Y., Lim, W. M., Jagani, K., & Kumar, S. (2023). Social media influencer marketing: Foundations, trends, and ways forward. *Electronic Commerce Research*, •••, 1–55.

Karwowski, M., & Raulinajtys-Grzybek, M. (2021). The application of corporate social responsibility (CSR) actions for mitigation of environmental, social, corporate governance (ESG) and reputational risk in integrated reports. *Corporate Social Responsibility and Environmental Management*, *28*(4), 1270–1284. DOI: 10.1002/csr.2137

Lazzeretti, L. (2023). What is the role of culture facing the digital revolution challenge? Some reflections for a research agenda. *Rethinking Culture and Creativity in the Digital Transformation*, 10-30.

Lescrauwaet, L., Wagner, H., Yoon, C., & Shukla, S. (2022). Adaptive legal frameworks and economic dynamics in emerging tech-nologies: Navigating the intersection for responsible innovation. *Law and Economics*, *16*(3), 202–220. DOI: 10.35335/laweco.v16i3.61

Liu, J., & Liu, Y. (2024). Marketing charm of magic kingdom: An analysis of Disneyland's marketing strategies. In *SHS Web of Conferences* (Vol. 193, p. 01008). EDP Sciences.

Maines, R. (2023). Gendered Capitalism: Sewing Machines and Multinational Business in Spain and Mexico, 1850–1940 by Paula A. de la Cruz-Fernández. *Technology and Culture*, *64*(4), 1303–1305. DOI: 10.1353/tech.2023.a911013

Mandracchia, C. M. (2021). *Transforming the Beast: The Theatre Laboratories of the "Disney Renaissance" 1984–1994* (Doctoral dissertation, University of Maryland, College Park).

Medoff, N. J., & Kaye, B. K. (2021). *Now media: The evolution of electronic communication*. Routledge. DOI: 10.4324/9781003020721

Moulton, C. (2022). *Having an Experience: Media Franchises, Events, and Participatory Culture* (Doctoral dissertation, Northwestern University).

Nyambane, R. (2021). The future of the printed book in the era of technological advancement: An imperative for digital innovation and engagement. *Journal of Information. Communication and Ethics in Society*, *19*(4), 537–559. DOI: 10.1108/JICES-10-2020-0106

Priadi, R., & Thariq, M. (2023). Reconstruction of Communicative Rationality: A Study on the Digital Mass Media Society in Indonesia. *Journal of Law and Sustainable Development*, *11*(4), e579–e579. DOI: 10.55908/sdgs.v11i4.579

Rachmad, Y. E. (2022). Revolutionizing Global Business: The Impact of Management 5.0 on International Corporate Strategies.

Raha, A., & Hajdini, I. (2023). Franchisees with multiple stakeholder roles: perceptions and conflict in franchise networks. In *Relationship Marketing in Franchising and Retailing* (pp. 48–67). Routledge. DOI: 10.4324/9781003364351-4

Rice, J., Raziq, M. M., Martin, N., Fieger, P., & Rice, B. (2023). The debt crisis and the adoption of Asset-Light and Fee-Orientated (ALFO) arrangements at Marriott: 1980-1995. *Journal of Tourism. Heritage & Services Marketing*, *9*(1), 58–66.

Sheoran, S., & Vij, S. (2023). A consumer-centric paradigm shift in business environment with the evolution of the Internet of Things: A literature review. *Vision (Basel)*, *27*(4), 431–442. DOI: 10.1177/09722629211033944

Tang, Q., Xie, E., & Reddy, K. S. (2022). Global production aspirations and internationalization by state-owned enterprises: A Co-evolutionary view of state-driven industrialized economy. *International Journal of Innovation Studies*, *6*(4), 276–285. DOI: 10.1016/j.ijis.2022.08.003

Tian, J. (2024). The Rise and Bypassing of Streaming Media. *Advances in Economics. Management and Political Sciences*, *57*, 189–195.

Tucker, E., & Bartkiw, T. J. (2022). Franchising and the extraction of surplus value: Excavating the legal boundary between franchisees and employees. *JL & Pol. Econ.*, *3*, 484.

Unnava, V., & Aravindakshan, A. (2021). How does consumer engagement evolve when brands post across multiple social media? *Journal of the Academy of Marketing Science*, *49*(5), 864–881. DOI: 10.1007/s11747-021-00785-z

Vassolo, R. S., Weisz, N., & Laker, B. (2024). The Digital revolution: harnessing the power of new technologies. In *Advanced Strategic Management: A Dynamic Approach to Competition* (pp. 257–273). Springer Nature Switzerland. DOI: 10.1007/978-3-031-57419-1_20

Williams, L. D. (2021). Concepts of Digital Economy and Industry 4.0 in Intelligent and information systems. *International Journal of Intelligent Networks*, *2*, 122–129. DOI: 10.1016/j.ijin.2021.09.002

Zhu, W. (2024). Decoding Disney's Marketing Mastery: A Strategic Analysis. In *SHS Web of Conferences* (Vol. 188, p. 03011). EDP Sciences. DOI: 10.1051/shsconf/202418803011

Section 4
Managing Technological Change Through Sustainable Business Models

Chapter 7
Future Challenges in Human Resource Management:
The Importance of Digital Skills and Corporate Social Responsibility

ABSTRACT

In the evolution of business management, ESG (environmental, social, governance) issues have an increasingly important role, which is no longer aimed exclusively at a reputational return but also above all at real ethical objectives and performance improvement. In this chapter it is emphasized the importance of these issues with particular attention to the social issue. The purpose is to demonstrate how the psychological well-being of the worker, which for decades has been put on the back burner, must be a primary objective of the company and must be seen not only as part of its outcome but as a real investment. It is precisely for this reason that it is of fundamental importance to demonstrate how, in addition to the obvious social benefit, correct management of human resources and their well-being can improve and speed up business processes, creating an effective and efficient working environment. The first part offers a comprehensive view of human resources management, first analyzing the history of HR management practice and then giving a clear vision of the role of the manager.

DOI: 10.4018/979-8-3373-3466-0.ch007

THE EVOLUTION OF HUMAN RESOURCE MANAGEMENT IN THE INDUSTRY 5.0 PARADIGM

Human resource management is a complex activity that encompasses all the practices that have to do with the relationship between the employed personnel and the company. In the person of the human resources manager, several processes are centralized, both the more specific ones regarding operational or marketing activities, and others aimed at a more general management of the value chain. The manager's work has a strong personal component because in addition to practical skills, he must be equipped with a strong emotional intelligence that enables him to make the best choices for his employees, while remaining aligned with the pre-established mission. It follows that each company has a specific policy regarding this topic and, even if the key principles are regulated or in any case universally shared, even small details of the relationship between the company and employees can have a fundamental impact, which is also why the topic of corporate engagement is increasingly discussed, it is in fact crucial to motivate the worker, and in the case of positions that are difficult to replace, to try to retain him in the company. This topic, which is now clear for all to see in the business landscape, has historically always been underestimated. This complexity that arose thanks to new laws and the growing presence of unions led to the birth of the figure of the Director of Personnel and Industrial Relations. This function will evolve in the following twenty years and will take the name of human resources, a centralization of the theme occurred precisely in this period with the looming of a problem of absolute relevance such as excessive corporate turnover, or a very high rate of employee turnover that brought a climate of chaos and disorganization to companies; today the term turnover is no longer associated with a threat and thanks to the evolution of human resources management, many companies see it as an opportunity for employees too, a statistic from LinkedIn (a leading social network for job hunting) has shown that in the first 10 years of their career, young people change jobs on average 4.2 times.

Nowadays, the practice of HR management has branched out into dozens of facets, leading the manager to reinvent himself also as an HR business partner and to have an increasingly important role in defining the company mission and the out as a product. The HR business partner has in fact a crucial and reference role for our thesis, as he is the figure who actively and materially connects the two hemispheres of business and welfare, having a role not only of supporting company decisions or of assisting the HR manager, but also of evaluating performance and monitoring the internal climate of the company, operating directly in the event of conflicts and/ or critical issues. He is the figure in charge of managing welfare plans, having the task of guaranteeing the best support to employees without losing efficiency and effectiveness in budget management. The HR Business partner must have a total vision of the company apparatus, which is why it is of fundamental importance that its projects are planned on a medium-long term, as short-term operations and decisions would not be able to make the organization stable and to accurately take into account all aspects of the company, maintaining the satisfaction of customers and employees at the highest possible level.

We have briefly analyzed the phases that involve exclusively the management of human resources, but a fundamental part of the relationship between company and workers is precisely in the sharing of values and objectives, which is why workers who embrace the values, mission and vision of the company represent an added value for it.

First of all, we must define these two terms: by mission we mean the reason why a specific company exists on the market, this term has a strictly operational connotation and indicates what the identity of the company is, how to achieve objectives, how to differentiate itself from competitors and what results to obtain.

By vision, on the other hand, we refer to long-term objectives, it defines what the purpose of the organization is and what problem it wants to solve for society.

Both are sides of the same coin, metaphorically speaking we can say that the vision indicates a destination, and the mission suggests the paths to follow.

They have a strong, attractive function, both towards potential investors and collaborators and towards the customers themselves; the vision is what makes the public passionate about the brand, while a solid mission is what allows the company to last overtime while remaining credible.

It is therefore essential to understand in which environment the company operates and what its corporate values are, some examples can be: diversity and inclusion, environmental sustainability, social impact and technological innovation, the issue of accountability is also important, that is, the coherence and tangibility of the actions that the company carries out, in order to avoid incurring cases of greenwashing.

Reconnecting to the scope of our thesis, these are fundamental elements to keep the worker motivated, as his degree of satisfaction does not depend only on the salary and the social position he holds, but on a more gratifying feeling of helping the world to progress through his skills and of being part of a healthy environment whose values he embraces.

This theme is increasingly recurring; several studies in fact state that one worker out of two does not love their job at all, this inevitably affects both the quality of life of the worker himself and his productivity.

When the word sustainability reaches the ears of a general public, it is inevitable that it is linked to issues concerning the environment and pollution, a correlation that is not incorrect but which only partially explains the concept of sustainability.

To better evaluate and define what it represents in Agenda 2030, ESG criteria are used, an acronym that stands for Environment, Social and Governance, which was discussed for the first time in 2004 when the then Secretary of the United Nations Kofi Annan had invited more than 50 CEOs of the main financial institutions to participate in a joint initiative that would integrate the values of the framework into the capital markets, the acronym will be coined a year later at the "Who cares wins" conference.

In the following decade, precisely in 2015, Italy joined 192 other countries in the UN initiative called Agenda 2030, which identified 17 SDGs (sustainability development goals) to be achieved, as the name suggests, by 2030.

Every year, AsviS (Italian Alliance for Sustainable Development) releases a report that includes an analysis of the progress of each goal and a comprehensive framework of policy recommendations to be submitted to the top to indicate the areas in which there is a need for action.

We pay particular attention to goal number 8 called "Decent work and economic growth" which, especially after the pandemic, is far from being achieved.

Remo Lucchi, President of the Eumetra advisory board, in an article for asvis, argues that increasing employee well-being can accelerate the achievement of the objective, having fundamental consequences: first of all, professional returns that are useful to the company, also for the employee's gratitude, but also the possibility of ensuring that employees perceive a significant increase in their well-being and are consequently encouraged to have a mindset more focused on sustainability issues, showing them a tangible improvement.

From an operational point of view, an ex ante analysis of the worker's status is suggested, both on a professional and private level, in order to calibrate targeted interventions to improve their condition; there must then be an ex post analysis with the identical research method to compare the before and after.

As mentioned, the pandemic has played a fundamental role in shifting attention to sustainability issues, it is precisely in the context of the recovery from covid 19 that the European Union has established the next generation EU fund, a recovery plan of 806.9 billion euros that has among its objectives a greener and more inclusive Europe with a particular focus on healthcare and investments in innovation and digitalization. It is precisely in this context that in Italy in 2021 the PNRR, National Recovery and Resilience Plan, was approved, divided into 7 missions:

- Digitalization, innovation, competitiveness, culture and tourism
- Green revolution and ecological transition
- Infrastructures for sustainable mobility
- Education and research
- Inclusion and cohesion

- Health
- Repower Eu (support for the ecological transition)

The PNRR has directly influenced the development of human resources management with targeted programs including:

The "National Program for the Employability Guarantee of Workers": a system for taking charge of the unemployed and people in employment transition, which uses tools such as profiling to create ad hoc paths for reintegration into the world of work.

The "National New Skills Plan": a plan focused on standardized training programs for the unemployed and on strengthening the training system.

The "New Skills Fund": a fund already tested in 2020 with the aim of reprogramming working hours in such a way as to replace hours of actual work with hours of paid training, based on specific agreements with trade unions.

The definition of a national gender equality certification system, which helps companies reduce the gender gap, with the aim of equal pay and job descriptions and greater protections and the introduction of gender difference management policies.

Strengthening the Apprenticeship Path and the Universal Civil Service

The aim is not to follow innovation but to guide it with specific training, support and change management programs, trying to instill a real cultural change in companies by demonstrating how adopting an ethical and sustainable path, strengthened with the help of modern technologies, can be the solution to the company's survival problems and the social problems that affect the world of work.

Recruitment, Selection, Training and Education

Recruitment corresponds to the moment where the company's job demand and job supply meet, it is therefore the first phase of the relationship between the company and the employee.

The modus operandi with which this phase is managed has evolved over time, as the needs of the company have evolved; in a more classic vision, only the previous qualities of the worker were taken into consideration, now much importance is also given to the ability to learn and therefore to the quality of human capital, this change is mainly due to a strengthening of the training and education phases, which allow the company to actively build talent.

The HR manager can make use of both internal and external recruitment tools.

Internal tools allow investing in resources that are already known, whose potential the company is fully aware of, there are both horizontal methods such as transfer to a different area and vertical methods such as the more classic promotion; a widely used technique is job-posting, which consists of informing employees of open positions and making public in the company panorama the requirements needed to fill a certain role.

Among the advantages of internal selection we have an increase in loyalty, a greater possibility of maintaining secrecy and a push for employees to work with greater enthusiasm and motivation, increasing productivity as each worker has the possibility of improving their condition without changing company.

External recruitment, on the other hand, offers the not insignificant advantage of having contacts with a very large pool of candidates, it can occur both directly and through diversified channels such as: advertising, employment agencies and self-nominations.

Even in this case, technology plays a fundamental role in supporting the management of human resources, in fact a widely used technique is e-recruiting, very popular in this sector are the social networks Indeed and LinkedIn, even if the use of more traditional social networks such as Facebook or even the reference websites of companies is not negligible.

The aforementioned employment agencies are instead divided between public and private and have the task of collecting the CVs of people in occupational transition to carry out matching operations.

Companies sometimes also carry out recruitment through schools, research institutes and universities.

The disadvantages that external recruitment presents are however different, as there are greater expenses, a greater possibility of running into profiles of no interest to the company and a limited knowledge of the previous experiences of the worker who is wanted to take charge, consequently these techniques are used for more complex and difficult to fill positions, when there is no figure in the internal panorama capable of covering a certain role.

Selection is the phase immediately following recruiting, it consists in choosing the subjects who will join the organization. Selection begins with a screening of the CVs received, based on generic criteria in order to make an initial selection based on level of education, linguistic and IT skills and work experience; this operation is sometimes outsourced by entrusting it to a selection company.

There are various selection techniques, some classic such as: interviews, professional tests and psychological tests that are often delegated to external bodies; other techniques are much more innovative, such as gamification, in which potential employees are subjected to games aimed at identifying enterprising and motivated resources, solving puzzles, or participating in contests, or competitions in which potential workers present projects to demonstrate their skills.

As in the recruitment phase, also in the selection phase the company can choose between the internal market and the external market, even if it is not completely free to do so since the selection is bound by the regulations in force in the country in which the company operates, such as laws in favor of equal opportunities, regulations on the protection of workers and any union agreements, we therefore note how already in this phase it is essential to maintain a balance between the well-being of the worker and the optimization of production. Selection costs are different and include: costs related to the collection of information and dissemination in the labor market, specific recruitment costs proportionate to the segments of the labor market and the number of candidates (recruitment cost per candidate and recruitment cost per hired), specific selection costs which, in addition to the same factors as recruitment costs, depend on the degree of refinement of the mechanisms chosen (selection cost per hired and selection cost per interviewee, costs of activating incoming and outgoing flows linked to remuneration and the adjustment between requested and offered remuneration) and finally the costs of conflict that arise when unions or stakeholders intervene in the process.

These costs and the associated risks often lead companies to prefer outsourcing choices also for this phase of resource management, these choices involve greater flexibility and a strong reduction in financial outlays, but sacrificing greater stability and lower transition costs.

Both the recruitment and selection phases are of fundamental importance, and the results of a correct management of the two phases are evident in the medium-long term; according to a study by McKinsey, strategically evaluating the different roles can increase the quality of hiring by 40% and reduce the dropout rate in the first year by 12%.

Training and Education

The training and education phases materialize in learning processes that favor the integration of human resources in the corporate landscape, in fact they have the purpose of adapting the skills of the workers to the needs of the company.

Although they refer to very similar contexts, the terms training and education have two distinct meanings: training is defined as the transfer of defined and controllable knowledge through didactic teachings or operational experiences (generally we talk about training in strictly operational functions; by training we mean everything that concerns the development of new skills or skills related to problem solving to deal with possible situations that have no historical precedents.

From this definition we can generalise by saying that there is usually a hierarchical distinction, we talk about training for figures with control, decision making and organisational roles; while training is often associated with the figure of the worker; however, even operational figures in recent years have had to carry out training processes due to the increase in the use of information technology in company processes.

The techniques used in training are different:

- Frontal lessons: often used for the transfer of models or information.
- The case method: a case is presented to the workers referring to a difficult situation that occurred in the company to test and train the problem solving skills of the employees, variations of this method are the incident where the learner must find the information and knowledge to solve the problem and the role playing in which the employees impersonate the actors of the company situation being examined.
- The in basket: a simulation technique that requires the participants to be subjected to different problems over the course of a day.
- Business Game: workers are divided into groups of companies that compete with each other.
- T. Group: participants are appropriately stimulated to analyze interpersonal relationships by improving communication and self-control.
- Learning communities, where multiple individuals collaborate through an exchange of skills that occurs autonomously.
- Action learning, in which participants are asked to create new projects in areas that do not belong to their area of operation.
- Group work, a classic method but which has countless advantages, such as the development of trust, the division of roles and the increase in social and communication skills.

In addition to these methods, the fundamental part of training is the so-called On the Job training, or experience in the field.

The training and education processes also include the evaluation of the skills transmitted, which is in fact extremely important to maximize the performance of the latter, as it must be remembered that each worker has a unique cultural background and predispositions to learning that must be valued to the best of their ability.

To carry out this evaluation, the process is as follows: it begins with learning tests, which are aimed at verifying that the student has learned the techniques taught, subsequently there is a phase of definition of the acceptability limits relating to the tests carried out, thanks to these limits we then have the parameters to carry out a correct verification of the effectiveness, a verification carried out in the field and over time for the evaluation of the skills learned. In addition to the effectiveness, the deviation is also calculated through an evaluation of the deviation of the acceptability parameters between field checks and documentary checks which helps to understand if there is the possibility of generating a non-conformity, which if correctly anticipated can be countered with a preventive action. If the two previous phases are not successful, non-conformity occurs, which translates into non-compliance with a defined and applicable requirement. Once a non-compliance has occurred, it is necessary to proceed with the analysis of the causes, that is, an assessment aimed at discovering the causes of the non-compliance that has the purpose of guiding a corrective action and preventing the problem from recurring in the future and intervening with a corrective action: an activity that serves to bring the requirement back into compliance.

We note how the company must assume the responsibility not only of choosing its employees in the best way but also of training them in the best way and in compliance with their skills and characteristics, this aspect nowadays is no longer negligible as each company has processes and procedures that are tailored to it and consequently it is no longer possible to imagine finding a "finished" worker.

EMPLOYEE WELLBEING

The term "employee wellbeing" refers to the health of the worker, both physical and psychological, and includes factors such as job satisfaction, or the employee satisfaction index, and work-life balance.

This is a very current topic, as it is common, especially among workers under 30, to accept job positions that do not allow them to have a healthy balance with their personal needs and that are often not consistent with the ambitions and type of career they would like to undertake.

Focusing on this topic brings countless advantages to the company, including: an increase in retention rates and productivity, a motivational boost for employees, a better experience for customers and a better brand reputation.

In an article for the well-known blog Zandesk, Hannah Wren identified five pillars to offer a total vision of what the topic of wellbeing in the workplace should be;

The first is Career well-being, or the possibility for the worker to grow and advance in their career, followed by Social well-being, which translates into the importance of the relational network within the workplace.

The third pillar is linked to the economic condition, or financial well-being: the company must guarantee the worker economic stability and competitive wages.

The last two: emotional well-being and physical well-being refer to the mental and physical health of the employee.

The history of worker well-being policies has origins that are far from recent, in fact we are talking about problems that date back to the industrial revolution, a period in which it was normal for women and men to work 18 to 20 hours a day, 7 days a week and it was equally normal to see children as young as 4 or 5 working in factories.

The first social revolution on this topic occurred in Great Britain thanks to Robert Marcus Owen who proposed a maximum of 10 working hours in 1810, in 1817 he suggested a further reduction from 10 to 8 hours, coining a slogan that still today gives a perfect definition of what the work life balance should be "eight hours labor, eight hours recreation, eight hours rest", Owen's ideals remained alive and were implemented in the following century, thanks to social activists around the world and companies that invested in change, among which the name Ford stands out.

The spread of these ideas went well beyond the Western world; in fact, during the years of colonialism by Great Britain in the mid-twentieth century, labor policies even reached India, which integrated them even after liberation, formalizing a treaty for 9 hours of work with a half hour rest and a maximum of 48 hours a week.

The First World War played a fundamental role, in which workers experienced 100-hour work weeks and the psychological effects were far from insignificant, there is talk of an impressive increase in the number of mental breakdowns and nervous breakdowns, Charles Myers, a reference figure for the study of the workforce, noted for the first time that problems such as monotony, lack of interest and repressed emotions were wiping out the productive capacity of employees.

The real change, however, occurred between the 70s and 80s where particular emphasis was placed on the quality of life, a leader in this sense was the company Johnson & Johnson, which in 1979 launched the Live for Life program.

In modern times, the biggest change was understanding that health was not only physical, in 2014 a report by Bucks Consultants showed an impressive figure: over half of the workers in the world have adopted strategies for the promotion of physical and mental health.

Another key aspect has been the reaction to covid-19, where workers have been subjected to unprecedented challenges such as forced isolation and the actual loss of opportunities for both personal and professional growth, all of which has brought wellbeing programs to another level by increasing the focus on individual solutions such as therapy and the offer of days off. In Italy this topic is so important that it has been introduced into the Civil Code, Article 2087 of the Civil Code states: "The entrepreneur is required to adopt, in the exercise of the business, the measures that, according to the particularity of the work, experience and technique, are necessary to protect the physical integrity and moral personality of the workers", this article detects both contractual and extra-contractual liability.

Other regulatory references issued in the last twenty years are:

- Legislative Decree no. 81/2008: it introduced the obligation to assess work-related stress risks.
- Legislative Decree no. 150/2009: this decree assigned the internal evaluation bodies (OIV) the task of conducting surveys on organizational well-being.
- Legislative Decree no. 33 /2013: it established the obligation to publish surveys on organizational well-being.
- Directive no. 3/2017 of the President of the Council of Ministers: contains guidelines for the implementation of paragraphs 1 and 2 of article 14 of law no. 24 of 7 August 2015 and the guidelines containing rules relating to the organization of work to promote the conciliation of employees' work-life balance.

Despite legislative support and growing awareness on the part of companies and members of the workforce, in Italy there are numerous situations in which workers are not respected and valued, this has serious consequences both on their career and on their private life; a study by the Observatory of the Polytechnic of Milan states that only 9% of workers believe they are happy with their position, a percentage that drops to 5% if we talk about employees who consider themselves happy with it.

The causes that generate this discontent are of various kinds, many are linked to salary and purchasing power, there are many cases of workers in large cities who, despite receiving an average salary, are unable to manage housing expenses, this highlights how important it is to analyze all the factors of the environment in which the company is located, because an equal salary in two different areas of Italy has a decidedly variable purchasing power. Another cause is to be found in the prospect of making a career and in the engagement of the worker himself, it is very common to feel that there must be a compromise between the work you are passionate about and the one that pays better, leaving the worker with a constant feeling of sacrifice. The time in the company and the distance from the workplace are other determining factors, an excessive amount of hours per week could not only damage the employee during the active hours of work, but also compromise the free time he has, as an excessively exploited worker will tend not to spend quality time once home but simply to rest while waiting for the next shift; help in this sense comes precisely thanks to technology that has allowed workers to access smart working days that allow them to be less constrained and to avoid having to bear the tiredness and costs that come from travel.

A very current theme regarding time in the workplace is working by objectives and not by hours, this according to several experts would improve motivation and reduce systematic absenteeism, as well as reducing the risk of having workers present but not very productive, we are therefore talking about MBO (Management by objectives), a term coined by Peter Drucker (1954) that refers to a system in which managers and collaborators decide which objectives must be achieved and in what time frame to achieve them, also encouraging workers to develop a personal time management that allows them to best match all aspects of their life.

Artificial Intelligence: A Help or a Potential Substitute?

Artificial intelligence is without a doubt the most important technological discovery of this century and although it is talked about a lot, the direct and indirect consequences it has on our everyday lives, and in particular on the world of work, are often not clear.

AI has a double implication, it is both a support and a potential substitute (Huang and Rust, 2018; Leone et al., 2021).

It is precisely the possibility that it will actually take away space from different categories of professionals that is frightening, generative AI is in fact capable not only of carrying out the most systematic tasks such as closing a balance sheet or creating statistics, but of replacing a real human effort, causing a crisis even for workers who make their intellectuality and uniqueness the main characteristic of their profession, even the art world has had to scale back as AI is able to give impressive results both in the music field and in the graphic and cinematographic fields, there is even talk of jobs such as dubbing that could disappear as they are easily replaceable, not only that, new categories of artists are emerging who express themselves by writing prompts to make the machine generate their works.

Considering that most artificial intelligences are available for free or at a low cost compared to their performance, which in addition to being effective is also immediate, it is not difficult to think of scenarios in which they carry out most of the known professions.

The very recent case of Klarna was striking, the Scandinavian company leader in the fintech sector has decided to replace half of its employees with artificial intelligence, according to the company's estimates their AI systems are able to perform the work of 700 human operators, reducing the average execution time of a task from 11 to 2 minutes.

This has inevitably led to a lower need for employees, in fact it has gone from 5000 to 3800, the CEO Sebastian Siemiatowski claims that the farewells are caused by spontaneous abandonments, in any case the company has increased the average turnover per employee by 73%, going from 4 to 7 million crowns (from 350 thousand to 620 thousand euros), also thanks to the downsizing.

This leads the CEO to promise a valorization of the staff who have remained within the company, arriving at paying individual employees much more, these optimistic forecasts are the result of the impressive increase in turnover in the first half of 2024 (in which AI systems were applied) which stands at 27%.

Artificial intelligence has significantly changed the work of the HR manager, it is used to support various phases, for example talent acquisition increasingly leverages AI for personnel research, in fact chatbots are used to evaluate candidates.

AI also allows for better acquisition and management of data and statistics, which allow talents to be trained using a much larger database and with the possibility of creating ad hoc paths, it also offers the possibility of showing the actual potential of workers and projecting it over the years within the company, creating an internal mobility strategy for talents and mapping the various career possibilities of the latter, improving productivity.

The adoption of AI also manages to compensate for a slight decline in investments in diversity, equity and inclusion, and therefore proves useful also from a social point of view; we can therefore say that hybrid work also helps to ensure culture, cohesion and responsibility of the staff.

The observatory of the Polytechnic of Milan has released some interesting statistics regarding the use of AI and the relationship that workers have with it:

> 26% of workers have used generative AI in the last year, but only 7% said they use it at least a couple of times a week;
> 24% of activities can already be carried out with the support of AI, almost one in two workers shows concern about the consequences: 12% are afraid of losing their job, while 26% are afraid of becoming more precarious, still 22% fear that their skills will become less relevant.

Contrary to pessimistic visions, there is a 29% of workers who believe that AI can be a valid ally to perform their work better, a 23% who believe that it helps them develop new skills and a 21% who feel lightened by the workload thanks to AI.

In conclusion on the topic we can say that the development of artificial intelligence is a fundamental part of progress, destined to change our everyday life, it is therefore essential to look at it as an opportunity and implement it in every business landscape, at the same time there is a need for social and ethical training at the base of this change, because such powerful tools in the hands of unprepared people or with incorrect intentions can generate irreparable social and environmental damage. It is therefore the task of modern business realities to ensure that it is an ally and a help for everyone, as it is the task of workers to update and adapt, but above all of institutions to create a solid system of rules and indications. Thus, AI is comparable to a modern and advanced version of the change that brought Harry Ford's assembly line.

CORPORATE SOCIAL RESPONSIBILITY (CSR)

The acronym CSR indicates Corporate social responsibility, or corporate social responsibility, which is an ethical approach to business management, aimed at satisfying all stakeholders, such as employees, suppliers and customers, as opposed to a traditional focus aimed at simply satisfying shareholders (Lindgreen and Swaen, 2010; Cezarino et al., 2022; Fatima and Elbanna, 2023).

It is precisely the attention to these aspects that forms the ethics of a company, which in modern times must have a social as well as economic function; both directly in the relationship with the various stakeholders, and indirectly through its choices: if, for example, a company agrees to collaborate with suppliers who underpay their employees to reduce costs, it is indirectly accepting an unfair practice and is also contributing to fueling a market tainted by competitive advantages that are not in line with progress in workers' rights.

It should not be understood as a cost, but as a long-term investment that has various benefits such as a strong impact on reputation: good CSR management helps to position the company, attracting not only investors, but also consumers themselves, who are increasingly aware and inclined to choose products and goods not only to their taste, but also in line with their ethical values. A clear example is the fight against fast fashion, which has dominated the fashion market in the last decade, but which now, despite still being a commercial practice used, in the common imagination is perceived as a "stain" on the company's reputation, so much so that even large companies such as Zara are led to carry out a rebranding aimed at moderately increasing costs, keeping them accessible, but at the same time maximizing the quality of the product.

In addition to attracting customers, reputation also serves to broaden the pool of potential investors and/or collaborators, who will always decide to place their brand alongside a company in line with their values. In fact, a good corporate image serves in the commercial field as a guarantee of stability. It should also be emphasized that there are risks associated with not using CSR. It is very likely that by not taking the ethical aspect into consideration, the company will be involved in scandals that would damage its image, and in modern times, damage to its image immediately translates into economic damage.

Good management of CSR generates a decidedly significant competitive advantage, because it also helps to optimize processes, reducing costs and improving efficiency.

Several studies have been carried out, both in Italy and internationally, on a correlation between CSR and financial performance. The Department of Management of the University of Bologna has in fact found a positive correlation between investments in CSR and the increase in revenues and profits; and even Harvard

Business School has shown how companies that adopt CSR practices are able to attract investments and reduce the cost of capital more easily.

The Osservatorio Socialis released a report in June 2022 that helps us understand how Italy has moved in the direction of Corporate Social Responsibility, among the most promising data we include:

- 2 billion and 162 million euros invested in CSR actions and in ESG areas.
- CSR is supported by 96% of companies with more than 80 employees, a sensational figure when compared to 2001 when the percentage stood at 44%.
- The positive trend has not been slowed down by the COVID-19 pandemic, with a 22% increase compared to the year the pandemic broke out
- The average investment in CSR by Italian companies in 2022 stood at 282,000 euros compared to 241,000 in 2021, an impressive increase in just one year, which confirms a trend that has been growing continuously for two decades now.

We can say with certainty that Italy has made, and continues to make, very important steps in the direction of CSR, this is certainly thanks to greater openness to innovation and dissemination by universities, research institutes and professionals; It must be remembered that when dealing with issues of this type, communication and accountability are always essential factors, as it is necessary to encourage a change in an ethical direction but also to give guarantees and demonstrate that the investment has a good return, in such a way as to convince even the most sceptical companies, which if not interested in the ethical side, can be attracted by a possible return.

The growth path towards a more sustainable world of work cannot depend exclusively on the decision-making processes carried out by companies, but needs to be guided and receive stable support from institutions, which have the duty to support progress and at the same time regulate the practices carried out.

Precisely for this reason, in recent decades, laws and guiding directives have been issued, both by the European Union and by the Italian State.

First of all, we must mention the CSRD: Corporate Sustainability Reporting Directive, Directive number 2022/2062 published in the EU Official Journal on 16 December 2022 as part of the European Green Deal.

The directive concerns the topic of corporate sustainability reporting, it has proposed to significantly increase the number of companies that will have to mandatorily draw up the Sustainability Report, i.e. a document that allows communicating company choices in the context of ESG criteria to customers, producers, investors, and employees.

The Directive extended the reporting requirement, which at the time was reserved for approximately 11,600 companies in Europe, to all large companies, both listed and unlisted, that exceed at least one of the following three limits:

- A number of employees greater than 250
- A balance sheet of no less than 20 million euros per year
- Net revenues of at least 40 million euros.

Microenterprises are excluded from this requirement, while small and medium-sized enterprises listed on the stock exchange are included.

For subsidiary companies with a non-EU parent company, net revenues greater than 150 million in the last two financial years must be considered, in addition they must have a subsidiary company that meets the CSRD requirements.

The Directive was issued with various objectives, such as making up for the shortcomings of the current legislation on non-financial information, innovating the communication system by companies and curbing the phenomenon of greenwashing, all this with the final objective of increasing clarity and transparency.

The sustainability report document must describe business models and strategies, highlighting risks and opportunities related to their choices, which must in turn be compatible with the transition objectives and the limitation of global warming to 1.5°C.

The response from the Italian State came with Legislative Decree no. 125 of 6 September 2024, which essentially takes up the European directive. It is no coincidence that it arrived more than a year and a half later than the CSRD of which it is a child. 2024 is the first year to be reported on 1 January 2025.

The Italian directive underlines the importance of describing the business model and the risk-opportunity ratio, the role and responsibilities of the administrative and control bodies, ESG policies, due diligence procedures with reference to the CSDD directive, and above all the negative impacts related to the company's activity, including the supply chain.

The directive does not lack important details such as the coherence between financial and sustainability information, it will be verified by an auditor and will initially be subject to a "limited Assurance" which then aims to evolve into a "reasonable Assurance"; to leave nothing to chance, the format in which it will be drawn up has also changed, namely XHTML, an innovation aimed at improving transparency and accessibility.

Worthy of mention is certainly the ISO 26000 standard, to introduce it we must first specify that ISO is an acronym that stands for International Organization for Standardization, it is the largest organization in the world that deals with defining technical standards in many areas, including that relating to management systems, and certifying that companies fall within the standards; ISO has produced over 22,000 standards since its inception.

Returning to the 26000 standard, it was released on November 1, 2010 and has as its theme Social Responsibility, it is not a certifiable Standard but rather a model that presents guidelines, suggestions and indications; the standard focuses on the identification and involvement of the various business areas on social responsibility and lists seven fundamental themes:

- Governance
- Human Rights
- Working relationships and conditions
- Environment
- Aspects related to consumers
- Internal area involvement
- Community development.

The peculiarity of this standard is that it can be extended to any type of company, even public.

In support of our thesis on the correlation between competitive advantages and sustainability, it is interesting to note that the standard in 2010 is a voluntary certification, and the companies that have adhered to it have found an optimization of internal processes, a reduction of risk factors and a significant increase in brand reputation.

The Role of Corporate Culture in Digital Innovation

A fundamental aspect that significantly influences a company's ability to innovate is its corporate culture. Corporate culture encompasses the values, norms, and behaviors that shape decision-making and interactions within an organization. A culture that embraces innovation fosters experimentation, collaboration, and openness to change—elements essential for successfully navigating digital transformation. On the other hand, a rigid and change-resistant environment can hinder the adoption of emerging technologies, slowing progress and limiting growth opportunities (Azeem et al., 2021).

Organizations that prioritize a culture of agility, adaptability, and openness to change are more likely to successfully implement innovative strategies and digital solutions. Leadership plays a crucial role in this process. Leaders who embrace new technologies and foster a mindset of change inspire employees to think creatively and challenge the status quo. They create an environment where employees feel empowered to contribute ideas and take ownership of innovation processes. This sense of autonomy and trust encourages teams to experiment with new digital solutions and take calculated risks.

Collaboration and knowledge sharing are also essential components of an innovation-driven corporate culture. Digital innovation thrives in environments where cross-functional teams work together, sharing insights and expertise. Breaking down silos between departments enhances the development and implementation of new technologies, leading to more seamless integration and better business outcomes. Additionally, companies that promote continuous learning and digital skills development ensure that their teams stay ahead of technological trends and can effectively implement digital tools. Organizations with a culture of lifelong learning are better positioned to adapt to market changes and sustain innovation over time (Ali et al., 2024).

A notable example of how corporate culture influences digital innovation is the case of Microsoft under the leadership of Satya Nadella. When Nadella took over as CEO in 2014, he inherited a company struggling with internal silos, resistance to external partnerships, and a culture that prioritized individual achievements over collective progress. Recognizing that technology alone was not enough to drive transformation, he focused on reshaping Microsoft's corporate culture. He introduced a "growth mindset" philosophy, encouraging employees to view challenges as learning opportunities rather than obstacles. This cultural shift enabled the company to embrace cloud computing, AI, and open-source collaborations—areas where it had previously been reluctant to engage. The result was a dramatic turnaround: Microsoft regained its competitive edge, surpassed $2 trillion in market capitalization, and became a leader in cloud services and AI-driven solutions (Francisco, 2024).

This case demonstrates that digital innovation is not solely a matter of adopting new technologies but also of fostering a culture that supports continuous learning, cross-functional collaboration, and adaptability. Companies that successfully integrate digital transformation into their culture encourage employees to experiment with new tools and embrace agile methodologies. They create environments where failure is not penalized but seen as a step toward innovation, allowing teams to iterate and refine ideas without fear of repercussions.

Moreover, corporate culture plays a pivotal role in managing resistance to change, one of the biggest barriers to digital transformation. Employees accustomed to traditional workflows may perceive digitalization as a threat to job security or an unnecessary disruption. Organizations that proactively address these concerns by fostering a transparent communication strategy and involving employees in the innovation process are more likely to succeed. Training programs, leadership support, and incentive structures aligned with innovation goals help reinforce a culture where digital transformation is viewed as an opportunity rather than a challenge (Naveed et al., 2024).

Ultimately, companies that prioritize culture alongside technology can sustain long-term digital innovation, ensuring they remain competitive in an increasingly fast-paced business landscape. Organizations that fail to recognize the cultural dimension of digital transformation risk investing in technology without fully realizing its potential, as resistance and misalignment can significantly undermine implementation efforts. By embedding a culture of continuous learning, adaptability, and collaboration, businesses can not only keep pace with technological advancements but also drive meaningful and sustainable innovation.

The Right to Disconnect

As we have already mentioned, society is constantly evolving and the development of new technologies, in addition to a variety of advantages, hides some pitfalls. Process innovation is essential to accelerate progress, and in modern times it is even necessary for it to take place, but new schemes and procedures bring with them new types of problems, both functional and ethical, that must be addressed.

This category includes a much-discussed topic in recent times, namely the problem of continuous availability that the worker unknowingly offers via cell phones and digital platforms, this means that the worker is continuously solicited, both directly by superiors or colleagues, and also by the information that he passively stores; this leads the worker to always feel like he has one foot in the workplace and never has the sensation of being free from commitments and duties.

This is why the right to disconnect was born, which is realized not only in the possibility of making oneself unavailable outside of actual working hours, but also in the freedom to do so without receiving sanctions, and in the right to be rewarded if the disconnection has not been exercised.

Being a very current issue, it still represents a legislative situation that is not perfectly defined, the development of which is underway at the time we are writing this thesis.

As regards the EU, on 21 January 2021 the European Parliament approved a Resolution inviting the European Commission to propose a Directive that would guarantee all workers the right to disconnect, underlining the negative consequences for workers from the point of view of physical and mental health, the EU urged member countries to introduce organizational measures to ensure this right that include both practical methods to allow the effective disconnection of the worker, and methods of assessing risks for health and safety; furthermore, innovations are foreseen in the systems for measuring the duration of working hours, awareness campaigns, training programs for workers and clear guidelines for when any derogations to this right can be granted, as each job has different existential conditions, in rescue work for example the availability is the basis of professionalism.

Duties have also been imposed for the Member States, in fact they undertake to guarantee the prohibition of discrimination, dismissal or other sanctions for workers who exercise the right, in this case it is the employer who is required to demonstrate that unfavorable treatment or dismissal is not retaliation; furthermore, the States must ensure that workers have access to a rapid, effective and impartial dispute resolution mechanism, as well as clear and comprehensive information on their rights provided by the employer in writing.

France was the first European country to regulate the right to disconnect with Loi n.2016-1088 of 8 August 2016 (Loi Travail or El Khomri), followed in 2017 by Spain and Italy.

It is interesting to note how each country has developed a different regulatory situation: some have chosen direct regulation, such as Greece, while others have entrusted it to collective bargaining, including Belgium, Ireland and Luxembourg.

The extent of this right varies according to the rules established, while in France the right applies to private companies with more than 50 employees, in Belgium it also extends to those with more than 20, while in countries such as Greece and Slovakia the right is recognized only for teleworkers.

The sanctioning systems are also significantly different from each other: in Spain, Portugal, Slovakia and Luxembourg there are in fact heavy financial sanctions for employers who do not respect the regulations on the right to disconnect, while at present this type of sanction does not exist in Italy.

The first regulatory reference in Italy is Law no. 81 of 22 May 2017, aimed at regulating agile work, this law does not establish a guaranteed and equal right to disconnect for everyone, unlike other EU countries, as its application is postponed to individual negotiations.

In particular, this legislation presents art. 18 which establishes that the organization of agile work is regulated by agreements between employer and employee and must respect the limits of maximum duration of daily and weekly working hours provided by the Law and collective bargaining, and art. 19 which specifies that the agreements must include references to the worker's rest times and the technical and organizational measures to be adopted to guarantee the disconnection of employees from the tech tools used for work.

In May 2020, the Privacy Guarantor expressed its opinion on the issue, stating that the non-recognition of the right to disconnect risks canceling the conquests obtained over the decades on work-life balance. In the same venue, the President of the Guarantor emphasized the need to prevent the monitoring that companies usually carry out on employees' activities through software that reports the actions carried out to superiors or blocks functions such as access to some applications.

The right to disconnect was then included in Law no. 61 of 6 May 2021, driven by the needs also arising from the complications resulting from the covid-19 pandemic.

This law establishes that workers who operate in smart working have the right to disconnect from technological tools and digital platforms in accordance with the agreements stipulated, taking into account any agreed periods of availability and without receiving negative repercussions or financial sanctions.

The importance of this issue was reiterated in the National Protocol on Agile Work in the private sector presented on 7 December 2021 and signed by the Ministry of Labour, the Protocol reiterated the importance of establishing disconnection time slots in which the worker does not provide the work performance and of ensuring that they are respected with technical and organizational measures.

At the time of writing this thesis, the DDL S. 1290 on the right to disconnect is being discussed in the Senate: the proposal applies to companies with more than 15 employees and states that the worker has the right not to receive communications from the employer or from the personnel invested with managerial duties towards the worker himself outside the working hours provided for by the applied contract and, in any case, for a minimum period of twelve hours from the end of the work shift.

Furthermore, administrative sanctions ranging from 500 to 3000 euros are intended to be introduced for each worker involved.

It is important to note that when it comes to employment contracts, history is cyclical, and even in the case of the right to disconnect, as in the case of working hours, individual agreements and company policies came before legislation. The Barilla case is very relevant in this sense, as it already excluded the possibility of contacting employees outside of smart-working hours in 2015. The positive implications for the worker who exercises the right to disconnect are multiple, thanks to it it is possible to prevent burnout and other psychological pathologies for the employee, in support of this the report "Right to disconnect: implementation and impact at company level" carried out by Eurofound in 2023 found that workers in companies without a policy on disconnection reported more frequently problems such as anxiety, depression, stress and migraines, while in companies where the right to disconnect is exercised there were high levels of professional satisfaction and a better work-life balance,

This gives us the opportunity to note how companies have a social as well as an economic function and that progress is achieved with active collaboration between institutions and private individuals.

IMPORTANCE OF EXCHANGING DIGITAL SKILLS FOR INNOVATION: MENTORING, REVERSE MENTORING AND LIFELONG LEARNING

The world of work is increasingly dynamic and the continuous change that the market requires often does not coincide with the ability to adapt of corporate organizational structures and employees themselves, for whom it is no longer enough to learn the basic tasks of their job, they in fact need to develop soft skills, such as adaptability, decision making and problem solving, or even more simply the ability to use new technologies.

While this may seem like a simple task for younger workers, who have a lot of room for adaptation and the possibility of learning, for older workers, accustomed to processes that have proven to be efficient over time but which may now prove to be decadent, it may be a difficult challenge.

It is precisely for this reason that the company must be able to respond to market needs and train workers, using methods that allow for a correct exchange of skills, which proves to be a competitive advantage for the company itself, and enhances the worker as it broadens his sphere of knowledge.

The need to innovate the technological systems of companies, and make employees able to use the new means in the most efficient way has generated the need for a new figure within the company apparatus: the Digital HR Manager.

This is a figure who has the function of preparing and training staff for the necessary digital renewal.

The role of the digital HR manager, although it may seem a strictly operational role, is actually very dependent on humanistic components of leadership and ability to motivate, his goal is in fact to trigger a behavioral change and in some ways also a cultural one, even the simple choice to eliminate the use of paper could in fact overturn consolidated habits in employees.

He has a fundamental role in supporting and making stable the system of use of Digital Platforms among workers.

When we talk about Digital Platforms we mean digital infrastructures capable of connecting the different systems together and exposing them to users through simplified and integrated interfaces, essentially often translated into mobile applications or websites.

They allow you to cross-reference and relate different data from different sources, in an accessible and effective way, with the advantage of being able to translate this information and simplify its understanding and visualization by users.

The peculiarity of digital platforms is the fact that individual services, generally focused only on a specific task, can communicate with other services belonging to the software and create an overall vision of the company's information and document flows.

These digital infrastructures are able to provide analysis, control dashboards and reports on data from different services, helping in the management and control phase; furthermore, thanks to them there is the possibility of creating customized interfaces for each type of user, this makes it easier for the worker to concentrate on their role, but at the same time allows them to access documents and information from other departments to verify their origin and analyze any critical issues.

The fuel of these digital platforms is undoubtedly represented by the databases that they have at their disposal, and the possibility that they have to manage them with a speed that cannot be replicated by a human being: it is a data economy based on knowledge.

The platforms are able to support not only the internal phases, but also the sales and post-sales phases, through them it is in fact possible to improve the customer experience and offer products more in line with their desires, just think of the analysis of cookies on websites that are able to provide important information on the interests of the potential buyer.

These data are provided voluntarily by consumers, and the possibility of obtaining them translates into an economic advantage for both parties: from the company's point of view there is the possibility of exponentially improving the effectiveness of the communication phase, as there is much more detailed information and an important screening can be carried out that allows not to waste advertising budget on a non-similar target, this allows the company the possibility of lowering costs and offering consumers important discounts.

The study “Banking on the platform economy”, which sampled 850 C-level companies in the financial sector, highlighted how 68% of consumers are willing to share their data with the bank in exchange for personalized offers, a percentage that drops to 46% in the insurance sector, 43% in telephone services and 41% in tourism.

We can say that nowadays these platforms are essential for the development and competitiveness of the company, they are in fact in support of some fundamental phases such as planning, human resources management and customer relationship management.

Companies that rely on digital platforms can customize them as they like, in order to best meet their needs, they are useful for any type of reality, whether it is a large financial company, a travel agency or even a simple restaurant, proving to be a great help in following the modern market characterized by increasing dynamism.

In addition to the innovative methods already mentioned for the exchange of skills, it is necessary to mention the classic practice of mentoring that has been used in small and large companies for decades, and which has origins well outside the corporate context, its name dates back to Greek mythology, Mentor was in fact a key character in the Homeric poem of the Odyssey, namely the friend of Ulysses charged with training his son Telemachus to become king of Ithaca.

Mentoring landed on the modern scene in the 1960s, when Yale University published a detailed analysis of the importance of mentors in supporting, guiding and orienting people to help them realize their careers, this study had such a resonance that it started a domino effect among American companies that began to integrate this practice in the decade immediately following.

Mentoring occurs when a highly experienced employee, usually older and higher up in the company hierarchy, who is called a mentor, provides guidance and support to a less experienced employee, called a "mentee" in English.

It is an extremely effective training system that has as its essential objectives helping employees to integrate and become productive, and encouraging them to develop their skills and improve at work.

The mentor can transmit to the mentee not only his operational knowledge, but also the values of the company itself, he can also inspire him by giving him a prospective vision of the role to which the mentee can aspire.

The selection phase of mentors is fundamental in the practice of mentoring, it is necessary to analyze their soft skills and choose those considered most similar, after which a training process is essential to make the mentor understand what his objectives are and how to communicate with the mentee, it must be emphasized that the relationship between mentor and mentee is a relationship between collaborators and consequently the traditional model of "professor-student" is not a winning model, consequently mentors must learn to manage communication without creating friction between colleagues.

Obviously, since mentors are employees just like mentees, they need to be encouraged to fulfill their role, usually two types of motivations are proposed: there is the material motivation, which is used mostly in rapidly growing companies, in which mentors (who often already have a more than significant workload) accept a monetary reward; otherwise we talk about immaterial motivations in which the mentor is given a status, for example direct gratitude from the manager or the use of certificates or badges.

Mentoring is a practice that dramatically accelerates the employee onboarding process, and if planned efficiently and properly controlled, it represents a significant source of attraction for new employees. Many Italian companies have started mentoring activities, among which Human Company, a company in the luxury tourism sector, and also Poste Italiane, stand out, which states that this practice is essential to generate networks, strengthen relationships and the overall vision while respecting diversity.

In the common view, the mentor is usually a more experienced employee, and generally older, than his mentee; but the speed with which the world of work is changing has also led this role to evolve, and has led new recruits, especially digital natives, to be fundamental in the process of training and professional updating of older colleagues.

Reverse mentoring therefore represents an inversion of the traditional mentoring system, based on a relationship of mutual learning and collaboration, in which the older person transmits his mindset and his work experience and the younger person transmits his innovative vision and soft skills that are certainly more dynamic.

The aim is to bridge the generation gap and break down the stereotype of the "professor" mentor who generates relationships of subordination of the student, the latter in fact presents the risk of creating a suffocating work environment for the employee, who would feel excessively controlled and not very free, it is therefore a type of practice that has to do with diversity management and inclusiveness.

Reverse mentoring was born almost at the same time as the web, in 1999, thanks to an intuition by Jack Welch, former CEO of General Electrics, for the need to maintain the experience of senior employees in the company and combine it with the ability to use new technologies of junior resources.

This practice has several advantages from a competitive point of view, in addition to those already mentioned in relation to inclusiveness there is also a clear strengthening of the leadership skills of the young person, a reduction in turnover and a general improvement in reputation.

A successful case of the use of reverse mentoring is certainly represented by the Barilla case, which to face the challenge of making the different generations of workers collaborate effectively created in 2017, as part of the company's Diversity & Inclusion strategy, the Y.O.U.N.G group (acronym for Young Original Unique Network Generations), it is an E.R.G (Employee Resource Group) that deals with generational issues.

It was the Y.O.U.N.G. group that promoted a path called "Reverse Mentoring – Generational Bridges" in 2018, relying on an external consultant with great experience such as Wise Grouwth.

The project started in 2018 with 12 people divided into 6 multigenerational couples, under the coordination of consultants Michele Rabaiotti and Alessia Alò. For this edition, the "train the trainers" method was used. The aim was essentially to train the trainers for subsequent editions. From the work carried out in 2018, a great collaborative ability was evident on the part of the selected workers and a general homogeneity of views, while a greater generational gap was also noted on the issues of sense of belonging and the approach to work.

The project was then extended in 2019 to another 30 colleagues from different company functions, renewing the structure of the project, which was then extended to the following year thanks to the encouraging results.

The project has seen the professional growth of ERG members and mentors who have been fundamental to the multi-year life of the project.

Reverse mentoring perfectly marries the philosophy of "lifelong learning", which extends the learning phase, usually relegated to the years of studies up to the first years of career, to the whole life, placing the emphasis on the need to be in continuous improvement.

The theme of lifelong education was born in an embryonic phase in 1930, when however it was closely linked to the education of workers, with a focus mainly on the operational part.

The first step towards the lifelong learning that we know today took place in 1972, when UNESCO released the Faure report entitled "learning to Be", where the importance of perpetual learning throughout life was essentially underlined.

The term learning replaced the word "education" in 1996 thanks to the publication of the Cresson White Paper, which made the term lifelong learning official and spoke for the first time of the Knowledge Society.

REFERENCES

Ali, M., Shujahat, M., Fatima, N., Lopes de Sousa Jabbour, A. B., Vo-Thanh, T., Salam, M. A., & Latan, H. (2024). Green HRM practices and corporate sustainability performance. *Management Decision*, *62*(11), 3681–3703. DOI: 10.1108/MD-05-2023-0787

Azeem, M., Ahmed, M., Haider, S., & Sajjad, M. (2021). Expanding competitive advantage through organizational culture, knowledge sharing and organizational innovation. *Technology in Society*, *66*, 101635. DOI: 10.1016/j.techsoc.2021.101635

Bastone, A., Leone, D., & Schiavone, F. (2023). Industrial Convergence and Digital Skills in Service Industries: An Explorative Analysis. *IEEE Transactions on Engineering Management*, *71*, 3354–3362. DOI: 10.1109/TEM.2023.3342641

Cezarino, L. O., Liboni, L. B., Hunter, T., Pacheco, L. M., & Martins, F. P. (2022). Corporate social responsibility in emerging markets: Opportunities and challenges for sustainability integration. *Journal of Cleaner Production*, *362*, 132224. DOI: 10.1016/j.jclepro.2022.132224

Fatima, T., & Elbanna, S. (2023). Corporate social responsibility (CSR) implementation: A review and a research agenda towards an integrative framework. *Journal of Business Ethics*, *183*(1), 105–121. DOI: 10.1007/s10551-022-05047-8 PMID: 35125567

Francisco, M. (2024). Correction: Artificial intelligence for environmental security: national, international, human and ecological perspectives (vol 61, 101250, 2023). *Current Opinion in Environmental Sustainability*, •••, 67.

Hannah Wren. (2024). " What is employee well-being and why is it important?", Zandesk, 16 aprile https://alleyoop.ilsole24ore.com/2024/03/13/4-days-working-week/ https://www.linkedin.com/pulse/tracing-origins-workplace-wellbeing-p6qof https://www.zendesk.com/it/blog/employee-wellbeing/

Huang, M. H., & Rust, R. T. (2018). Artificial intelligence in service. *Journal of Service Research*, *21*(2), 155–172. DOI: 10.1177/1094670517752459

Leone, D., Schiavone, F., Appio, F. P., & Chiao, B. (2021). How does artificial intelligence enable and enhance value co-creation in industrial markets? An exploratory case study in the healthcare ecosystem. *Journal of Business Research*, *129*, 849–859. DOI: 10.1016/j.jbusres.2020.11.008

Lindgreen, A., & Swaen, V. (2010). Corporate social responsibility. *International Journal of Management Reviews*, *12*(1), 1–7. DOI: 10.1111/j.1468-2370.2009.00277.x

Naveed, M., Akhtar, S., & Siraj, D. (2025). Navigating the Digital Shift: Analyzing Teachers' Perceptions of Transitioning to Online Platforms at Higher Education. *iRASD. The Journal of Educational Research*, *6*(1), 10–21.

Peter, F. (1954). Drucker. *The Practice of Management*, 72.

Schiavone, F., Leone, D., Caporuscio, A., & Kumar, A. (2022). Revealing the role of intellectual capital in digitalized health networks. A meso-level analysis for building and monitoring a KPI dashboard. *Technological Forecasting and Social Change*, *175*, 121325. DOI: 10.1016/j.techfore.2021.121325

Chapter 8
Technological Change Management for Sustainability

ABSTRACT

This final chapter aims to open the path for future research in the literature on technological change management. Innovative strategic approaches, new technologies, and new business models for achieving sustainability were discussed in-depth, but doubts remain about the actual applications within business processes to avoid chaos in organizations. This led the author to ask himself if they were sustainable economically by improving the attractiveness of some businesses and environmentally by enhancing the efficiency of how raw materials are used, but from a social point of view. This change is also verified in the production sector and has then inevitably extended to an economic-social context, modifying the way in which work is organized and influencing consumer choices. It is, therefore, a change that has occurred progressively and is supported by digital innovations that have made industrial production increasingly compliant with market demands. It was analyzed whether there had been an improvement or worsening of the working conditions of employees or compliance.

SUSTAINABILITY, GREEN ECONOMY, AND THE NEED FOR A LEAN APPROACH

> *"When we talk about sustainable development, we must not limit ourselves to considering only environmental issues, but we must also take into account the economic and social sustainability of activities and investments" (McQuade and Schmitz, 2017).*

DOI: 10.4018/979-8-3373-3466-0.ch008

The concept of sustainable development must be analyzed in terms of: efficiency, equity and compatibility. To be understood as a continuous and necessary process, to unite the three fundamental dimensions for environmental, economic and socio-cultural development. From an economic point of view, sustainability translates into a long-term approach, where financial interests are balanced with social and environmental ones. This involves targeted investments, social impact assessments and corporate ethics.

In the context of studies on corporate social responsibility and sustainable development, it is appropriate to refer to the studies conducted in the early 1990s by John Elkington who summarized the idea of a sustainable business in the term "Triple Bottom Line", a business approach model that focuses on three main areas, introduced specifically to encourage companies to evaluate their performance not only in terms of profit, but also in relation to social and environmental impact. The company that, in respect of individual, collective and environmental well-being, seeks suitable levels of profitability, pursues conditions of corporate sustainability.

Concern for the environment is now seen as a business best practice, due to pressures from various sources. It has become common for companies to use the concept of climate change, or more precisely the actions taken to prevent it, as a marketing tool. From the use of renewable energy sources in paper production to hybrid cars, the business sector seems to be responding to public concerns about the global climate.

Ans Kolk and Jonatan Pinkse (2007), in their article "Towards strategic stakeholder management? Integrating perspectives on sustainability challenges such as corporate responses to climate change", examine how the impact of different stakeholders is changing marketing, production values and corporate best practices regarding environmental impact, the authors describe stakeholders as "any group or individual who can influence or is influenced by the achievement of the organization's objectives".

Government regulation plays an important role in determining the level of commitment that companies give to environmental issues, as regulations and laws have a strong direct impact on the prevention of pollution. In the event that the legal system does not impose limits on a company's emissions, non-governmental organizations (NGOs) exert pressure, cooperating to control the amount of carbon emissions.

Geographical location also plays a significant role in determining the degree of commitment of companies to pursue and implement emissions reduction targets. Therefore, companies operating in countries that have ratified the Kyoto Protocol tend to obtain more effective measures to reduce their impact on the environment. In these countries, regulatory requirements regarding business practices are established, that is, organizational policies aimed at reducing greenhouse gas emissions are implemented more frequently than in countries that rely on non-governmental organizations.

The increase in the intensity of production processes, economic growth and the expansion of urban settlements have accentuated the pressure on environmental degradation.

The traditional model of industrial market development was challenged and questioned, when in the 1960s, the recognition of phenomena, such as environmental problems and the effects of climate change, led to the birth of the concept of eco-development and "green economy"; "The concept of green economy focuses on the relationship and interdependence between economy and ecosystem, providing a basis for operationalizing the concept of sustainability".

Going back, to better understand this phenomenon, it is necessary to refer to the studies conducted by Herman E. Daly, one of the founders of ecological economics.

Since the 1960s, the industry has undergone a significant evolution thanks to the introduction of the first digital technologies, which have gradually replaced production based on mechanical technologies. This transformation marks the transition from mass production to a more flexible model, the so-called Lean Production.

In this more "lean" production approach, the volume and quality of orders determine the pace of production: companies regulate both the flow of raw materials, with the aim of minimizing warehouse stock and operating costs, and the workforce, introducing the Just in Time production model. The basis of this latter model is a production divided into a series of independent micro-activities and coordination between all the actors involved, supported by the most advanced use of digital technologies.

This change in the production field has then inevitably extended to an economic-social context, modifying the way in which work is organized and influencing consumer choices. It is therefore a change that occurred progressively and supported by digital innovations that have made industrial production increasingly compliant with market demands.

After 2000, the phenomenon of digitalization began to exploit advanced tools for the dissemination of codified knowledge, using the global network to share information in real time and at extremely low costs, if any. Thanks to this type of technology, large companies are able to generate economic value by developing innovative standards capable of reusing digital codes and algorithms up to a million times. This process avoids mass production (typical of Fordism) and allows physical solutions to be adapted to increasingly complex contexts, for example thanks to machine learning processes. It is therefore clear that a first phase of digitalization was characterized by the global diffusion of innovative standards, while a second phase is oriented towards the customization of products and services, based on the specific needs of individual users. Technology, as essential to achieving a competitive advantage, is present in every phase of the value chain, since all activities require technological skills in addition to market skills. Although the technological dimension is widespread in different areas, it is essential to identify an "internal sector" of the company in which these skills are generated. This sector can be represented by the technological development activity, which goes beyond the mere research and development activity, including the set of activities and knowledge aimed at improving and innovating products and processes. Consequently, the costs associated with this activity derive from: research and development, acquisition of know-how, implementation of information systems, production technologies and related studies and projects.

Therefore, technological development is placed among the support activities of the value chain, significantly influencing all the primary phases of the chain.

Smart Manufacturing is one of the major evolutions of modern industry, radically transforming the production landscape by using internet-connected machinery to monitor and optimize operations. The main goal of this innovation is to increase operational efficiency and productivity, identify automation opportunities, and exploit real-time data. Within factories and along the supply chain, smart manufacturing facilitates the integration of physical and digital processes. It also enhances decision-making by transforming interactions between people, processes, and technology, thus ensuring greater efficiency and agility. Artificial intelligence plays a crucial role in this context, analyzing the vast flow of data to optimize production processes, ensure the quality of goods produced, and even predict equipment failures, thus enabling the scheduling of preventive maintenance. Despite its potential, however, the adoption of SM faces significant obstacles, such as the lack of standardized technical protocols that limit communication and interoperability between different systems. In addition, there is the problem of limited accessibility for small businesses due to high initial costs and technological complexity.

However, with the progress of MS, an increasing integration of automation, AI and other technological innovations, such as digital twins (virtual replicas of a physical resource), is expected for a future characterized by hyper-connected and intelligent production.

According to a study conducted by some researchers from INAIL, today technological evolution is closely linked to work safety systems. The attention of researchers has turned to the impact that Industry 4.0 has on worker safety, particularly in the manufacturing sector; safety, in fact, is considered an essential actor of innovation. One of the examples analyzed concerns the adoption of a system for preventing visual injuries, which have significantly decreased thanks to augmented reality technology, which allows the user to interact through visors and sensors, allowing a realistic perception of three-dimensional environments. In this way, the worker not only has the opportunity to familiarize himself with equipment and environments, but also manages to receive training on risk situations without an actual danger to his safety. Smart manufacturing can also be seen as a synonym for servitization: a service-focused approach that integrates the complex ecosystem of supply chain players. This allows them to collaborate synergistically to ensure maximum customer satisfaction and loyalty. At the center of the new industrial model today is the concept of Industry 5.0, born in 2015 as a paradigm focused not so much on technology, but on a cultural revolution, which places attention on: an eco-sustainable approach in the production phases and on the well-being of workers. In this perspective, technology is used to benefit the quality of life of the human being by improving human capabilities, but not replacing them.

The birth of this competence derives from the combination of material resources, such as equipment, and intangible resources, such as patents, tacit knowledge or human capital. In this regard, technological development can also become a distinctive competence of the company when it contributes to generating a competitive advantage, whether we are talking about cost leadership or differentiation, thus favoring the creation of uniqueness within the company.

Generally, the technological development phase is preceded by the Research and Development (R&D) phase as a starting point for innovation which, through studies and research, allows the creation or improvement of new technologies. R&D therefore allows the development of efficient production methods by improving the competitiveness of the technology developed.

Today, the focus of Project Management is no longer just productivity, but also the ability of teams to adapt to change, thus determining the possibility that investments in R&D translate into concrete results. As a result, R&D strategies have become the source for the development of strategies for new business models.

SUSTAINABLE BUSINESS MODELS IN ORGANIZATIONS AND THE NEW SOCIETY 5.0

A company is defined as sustainable if it embraces and implements the principles of equally sustainable development. Therefore, it determines its choices, activities, services or products offered through full awareness of the economic, environmental and social consequences, ensuring that it has a positive or at least neutral impact on the various categories of stakeholders.

After having studied the concepts of business model innovation and sustainable business model, the case of sustainable business model innovation will be analyzed below. The difference between the concepts of sustainable business model and sustainable business model innovation is very subtle, the first refers to a company in which there is a logic of sustainability within its business model, while the second regards the process by which the company innovates its current business model through the tool of sustainability. The sustainable business model literature often refers to the development of a stable model, while it is more interesting to study the adaptation and transformation of a business model. In fact, sustainable development requires continuous mutations to allow the company to continuously adapt to the changing environmental context (Roome and Louche, 2016). By developing the ability to adapt, the organization becomes resilient. This characteristic then allows her to face problems and seize the opportunities that present themselves over time, managing to be long-lived.

As described in the previous paragraphs, the theoretical evidence seems to suggest that innovation is the key to increasing company performance. Innovation through sustainability could allow companies to seize many interesting business opportunities and at the same time improve environmental and social conditions. It is for this reason that the sustainable business model innovation represents today, according to many scholars, the most interesting innovative form. What is not clear in the literature is the link between sustainable innovation and performance.

Empirical evidence demonstrates that the two concepts have a positive correlation: it can be observed that sustainable business model innovation is often accompanied by an increase in the performance of the organization that successfully adopts it; however, it is not clear how and with what intensity this business structure affects performance. To better understand this issue, an analysis of the literature will be carried out below. As can be easily understood, sustainable business model innovation merges the theory of business model innovation with the sustainable business model. Business model innovation is a corporate transformation process, to qualify it as sustainable business model innovation, this process must be oriented towards:

- Promote sustainable development by increasing positive impacts or decreasing negative ones on the environment and society. The organization must adopt a long-term perspective and involve as many stakeholders as possible in its action.
- Incorporate sustainable logic into your business model by reviewing the value proposition, value creation, value capture and value networks (Geissdoerfer et al., 2018).

Furthermore, as Geissdoerfer et al., (2018) illustrate, it is possible to distinguish four types of sustainable business model innovation:

- Sustainable start-up: a new organization adopting a sustainable business model.

- Sustainable business model transformation: the company transforms its traditional business model into a sustainable business model.
- Sustainable business model diversification: the company does not make major changes in its business model but adds a new sustainable business model to the existing one.
- Sustainable business model acquisition: the company identifies an interesting sustainable business model and then proceeds to acquire it and integrate it into the existing organization. These four types of sustainable business model innovation aim to implement certain types of sustainable business models and strategies. These strategies, identified by various scholars, were then reviewed by Bocken et al. (2014) and updated by Ritala et al. (2018).

These scholars have identified nine types of generic sustainable business models, called "archetypes", the strategies adopted are as follows:

Maximize material and energy efficiency
Close resource loops
Use renewable sources and natural processes
Offer features rather than properties
Adopt a guardianship role
Encourage sufficiency
Reuse for society or the environment
Create inclusive value
Develop sustainable solutions at scale

According to Lüdeke and Freund (2010) business model innovation is increasingly recognized as the key to providing greater social and environmental sustainability in the industrial system (Bocken et al., 2014). Sustainability-related challenges can be seen as opportunities for innovation for organizations. To be able to take full advantage of these opportunities, organizations must be able to solve problems in creative ways. In a process of Schumpeterian destructive creation i sustainable entrepreneurs destroy traditional production methods and their business models by replacing them with more sustainable (or less unsustainable) business models (Schaltegger et al., 2018).

In doing so they combine two types of objectives: to contribute to more sustainable development while discovering, evaluating and exploiting new economic opportunities (Dean and McMullen, 2007). Based on the above considerations, Schaltegger et al. (2016) define sustainable entrepreneurship as: "A mission-driven sustainability process to solve environmental and social problems of unsustainability by exploring and exploiting market opportunities created through innovative business models". The definition used here for sustainable business model innovation is the one described by Geissdoerfer et al. (2018): " Sustainable business model innovation concerns the development and implementation of a sustainable business model within a firm [...]" Very interesting, regarding the analysis of sustainable business model innovation is the report between sustainability and financial performance.

It is important to try to better understand this issue, because even if environmental and social issues are very important, companies often think, in the first place, strongly considering the economic sphere. For this reason, finding a positive correlation between environmental and social improvement and economic improvement, understanding how and to what extent the former influences the latter, could lead managers who are still hesitant about the subject to adopt a more sustainable business model without delay. Unfortunately, doing a review of the literature it is possible to observe that the relationship "sustainability - economic performance" has not been treated in a very detailed way. The reason for this can be traced back to the difficulty of estimating the effects of innovative practices in the sustainable field with the increase in financial performance.

For this reason this thesis will try to fill the gap in the literature, in the hope of stimulating new theoretical research on the subject. Below we will see the most important considerations listed by scholars in this area. In chapter III, through an empirical study we will try to better understand the topic under examination. Initially it was assumed that companies' attention to sustainability had a negative impact on performance, due to the higher costs that the implementation of this logic would have entailed. Subsequently, through many studies and empirical research, it has been demonstrated that sustainability is positively related to the increase in competitiveness of a company.

Despite this, it is not easy to identify the positive impacts of sustainable business model innovation on performance, also because current measurement tools make analysis difficult, especially as regards environmental and social aspects. Furthermore, speaking of sustainability, the creation of value is not limited to the company's perspective (creation of value for the stakeholders), but the horizon must be broadened to consider a vast number of stakeholders influenced by the company's activity (Evans et al., 2017).

For simplicity, this analysis will omit the value created for the various stakeholders and will concentrate exclusively on the value created for the shareholders, ie the economic value. As mentioned earlier, growing global competition has prompted companies to look for new opportunities to stay competitive. One of these is represented by creating new markets through the tool of innovation. Innovation in the sustainable field is often traced back to a win-win situation (Kramer and Porter, 2011), i.e. the increase in economic performance and more sustainable development are pursued at the same time, for this reason, even if it is not possible to precisely establish the impact of sustainable business model innovation on performance, companies should take steps to implement this process.

By applying this new business logic, they could appreciate significant all-round improvements through the construction of a sufficiently flexible and future-oriented structure, capable of seizing opportunities and countering the threats that will arise. If the organization can successfully apply all of this, it can be resilient. Resilience is the company's ability to react to external/internal shocks by fueling a virtuous process of continuous transformation and adaptation. Only companies that will be able to develop this feature will be able to be long-lived.

Furthermore, sustainability can take on different facets in the corporate system; in fact, the degree of sustainability depends on the level of penetration of sustainable principles, the importance given to stakeholder engagement, and the social, economic, and environmental impact produced.

Sustainable development materializes in the pursuit of sustainability along three closely interrelated dimensions: economic, social and environmental dimensions.

In 1997 John Elkington, in the article "Partnerships from cannibals with Forks: the triple bottom line of 21st-century business", underlined this concept and with the expression "the triple bottom line" Elkington intended to promote the measurement of the performance of a company in relation not only to the economic result but also to the social and environmental results; therefore, the sustainable development model includes the constant improvement of business performance along the three dimensions.

In the article Elkington identified three aspects on which to direct the company's focus in line with the TBL (Task-Based Learning) approach, i.e. the approach based on the performance of targeted tasks: profit, people and Planet:

- Profit: is the traditional measure for evaluating the economic performance of the company. The basic assumption in this aspect is simple: if the company is not capable of producing economic profits it is not destined to survive over time. It is useful to point out that profits mean the creation of value in the long term;
- People: it is important for the company to evaluate its impact on people and, therefore, on society which it interacts with. In this regard, the socially responsible corporate figure is promoted;
- Planet: concerns the impact that the company has on the environment.

The first pillar is that of environmental sustainability.

Figure 1. Comparison framework for Industry 5.0 and Society 5.0. (Adapted from Huang et al., 2022).

When we talk about "environmental sustainability" we are referring to the natural resources and biophysical conditions made available to man by our planet and to the ways of conscious use, that is, they cannot be used, exploited and compromised indefinitely without thinking about generations futures; therefore, use them in a sustainable way.

Recently, during the pandemic, we have witnessed poor management of resources and the environment, the management of which has caused enormous damage on a social and economic level. They can be considered, for example, the natural disasters that occurred in Germany or the water bombs last summer. These phenomena are the result of global warming. During the lockdown we noticed how environmental conditions improved given the decline in the use of petrol cars and planes.

In order to be defined as sustainable, a system must have the prerogative that its consumption is remodulated, recycled, reusable and biodegradable; in fact, a sustainable system is defined as that which is capable of self-regulating and self-defending from internal and external agents that could alter its natural balance.

A non-profit organization "The natural Step" has pointed out that environmental sustainability concerns:

- Reduction in the extraction of natural substances from the earth's crust;
- The reduction of the production of chemical substances and compounds;
- The reduction of physical degradation of the nature of natural processes;
- The reduction of obstacles that do not allow people to satisfy basic human needs.

The second pillar of sustainability is represented by "social sustainability", which consists in the ability to maintain a condition of well-being distributed equally within society.

The term "well-being" includes a multiplicity of concepts such as security, health, education, justice and democracy. From this point of view, our country unfortunately does not have a leading position.

Following the financial crisis of 2007/2008, Italian education, healthcare and research suffered significant cuts. Furthermore, there are inequalities between the various social groups of the population which do not allow the creation of a sustainable development model.

Over the years, numerous initiatives for social sustainability have been included, in fact, many companies have given increasingly more importance to the social aspect of the concept of sustainability, introducing Corporate Social Responsibility to launch corporate sustainability and welfare projects.

Among the various initiatives there are:

- Company nurseries: a type of welfare that helps families reduce costs and times linked to family dynamics;
- The neighborhood economy: is an initiative that supports small neighborhood shops to encourage territorial development and local production;

The last pillar, but not least, is economic sustainability, because in every dimension sustainable development is contained in the interest of the economy. It can be defined in practical terms as the production and maintenance of favorable conditions for economic growth through the exploitation of resources such as labour, capital and territory which allows the long-term preservation of the economy's capacity for action, thus being able to enhance the products and services of the territory, increase the GDP and also reduce unemployment.

In order to define a "sustainable" economic model, resources must be exploited slowly compared to how much they are reproduced and the focus on external variables such as education, population growth and research is of particular importance.

In this regard, a specific economic model was created to represent a valid alternative to the Linear Economy, which allows the concept of economic growth to be separated from the consumption of resources: the Circular Economy. the Circular Economy allows us to re-propose production and consumption models to reduce waste and reuse materials within infinite production cycles.

The circular economy is characterized by the regeneration of products and components, capable of eliminating the concept of the end of a product's life and bringing significant environmental benefits. Thanks to the circular economy it is, in fact, possible to give waste a second life.

Looking at the industrial paradigms, Industry 4.0 represents a significant transformation of the industrial landscape by integrating advanced technologies such as the Internet of Things (IoT), Artificial Intelligence (AI), robotics, big data, and augmented reality. However, despite its many benefits, Industry 5.0 has several shortcomings that must be addressed to fully exploit its potential and ensure sustainable and inclusive development.

Increasing connectivity and digitalisation brings new challenges in terms of cybersecurity and data protection.

¬ Cyber Threats: The proliferation of connected devices increases attack surfaces, making businesses more vulnerable to cyberattacks that can disrupt operations, compromise data security and cause significant financial and reputational damage.

¬ Regulatory Compliance: Data protection regulations, such as the GDPR in Europe, impose strict requirements for the management and protection of personal data. Compliance with these regulations can increase complexity and operational costs for businesses, requiring significant investments in security infrastructure and staff training.

Advanced automation and the use of intelligent robots can lead to the replacement of human jobs, causing technological unemployment.

¬ Job Loss: Automation can eliminate manual and repetitive jobs, creating a disparity between supply and demand for jobs. This can lead to increased unemployment and social tensions, especially among less skilled workers.

¬ Skills Gap: The rapid evolution of technologies requires new skills and qualifications that not all workers possess. Businesses need to invest in staff training and reskilling to fill this gap, but this requires significant time and resources.

The implementation of Industry 4.0 technologies also raises ethical and social questions.

¬ Data Privacy: The massive collection of personal data raises concerns about privacy and the ethical use of information. Businesses must be transparent about how they use data and must ensure that consumer data is protected from abuse.
¬ Inequalities: Access to advanced technologies can be uneven, favoring large businesses and developed economies to the detriment of SMEs and developing countries. This can increase economic and social inequalities, creating a technological gap between those who can afford to invest in Industry 4.0 and those who cannot.

The adoption of Industry 4.0 requires advanced technological infrastructures that are not always available.

¬ Connectivity: In many regions of the world, fast and reliable Internet connectivity is not yet a reality. This limits businesses' ability to adopt IoT technologies and connect their systems effectively.
¬ Access to Funds: The infrastructure needed to implement Industry 4.0 requires significant investments. SMEs often struggle to access the funds needed to upgrade their technology infrastructure, falling behind larger, better-financed competitors.

Despite the many benefits of Industry 4.0, it is essential to address these shortcomings to ensure that the transition to industrial digitalisation is sustainable, inclusive and safe. Government policies, collaboration between industry and academia, and the commitment of companies towards continuous training and responsible innovation are crucial to overcoming these challenges and realizing the full potential of Industry 5.0.

Due to the shocks that have hit the world (Covid-19, Ukrainian conflict, environmental crisis), Industry 4.0 is no longer able to meet the objectives required by society. Given changing lifestyle habits and new needs, the industry requires a redesign, focusing on new objectives. It is natural that an exclusively technical paradigm, focused on efficiency through connectivity and artificial intelligence, can no longer be considered adequate. What we observe is that the economic dimension cannot be the exclusive consideration. In fact, as stated in the 2030 Agenda, the Sustainable Development Goals have signaled the need for a course correction in Europe. This is a 2015 action plan pushed by the United Nations (UN) that aims to promote people, prosperity and the environment. Following the results of the Millennium Development Goals that preceded them, and represent common objectives: that is, they concern all countries and all individuals without leaving anyone behind along the path necessary to bring the world on road to sustainability.

On the contrary, Industry 5.0 results in a "winner-takes-all" model that generates monopolies and gigantic technological inequalities. Taking into consideration the 17 points described in the 2030 Agenda, it is possible to identify how the new industrial strategy should be designed. The shortcomings of Industry 4.0, according to Sandrine Dixson-Declève (sustainability scholar), can be summarized in three dimensions:

1) The regenerative characteristics of industrial transformation, so as to embrace the circular economy and positive feedback loops, not as a afterthought, but as a fundamental pillar of the design of entire projects.
2) An intrinsically social dimension, which requires attention to the well-being of workers, the need for social inclusion and the adoption of technologies that do not replace, but rather complement human capabilities, when possible.
3) A mandatory environmental dimension, which leads to promoting a transformation that eliminates the use of fossil fuels, promotes energy efficiency, draws on nature-based solutions, regenerates carbon sinks, restores biodiversity and creates new ways to thrive in respectful interdependence with natural systems.

As a result, the European Union sees the need for a long-term industrial strategy that focuses on the digital and low-carbon economy, enabling businesses to remain competitive. This new perspective is known as Industry 5.0. 3. Society 5.0. Japan played a fundamental role in defining a new social vision. In fact, even the already described Industry 4.0 has been conceptualized differently, identified as Super Smart Society. It is crucial to underline that Industry 4.0 is not the only component that can be recommended for society's paradigm shifts. The new vision of society reflects some of the ideals introduced by the United Nations with the Sustainable Development Goals plan.

The term Society 5.0 was announced for the first time by the Japanese Government, during the Fifth Basic Science and Technology Plan. Society 5.0 is a human-centered society, with the aim of satisfying personal needs by promoting economic and social equality, facilitating the fusion between cyber space and physical space, thus generating an improvement in data collection, which enriches the problem solving and value creation. There are two essential components of the transition to Society 5.0, namely managerial thinking and Japanese culture. In fact, managerial culture and practices illuminate the necessary collaboration between people and institutions, creating less formal relationships. The role of the individual is crucial, the centrality of man within the new society is a handbook for understanding society 5.0. While, from a management point of view, the correlation between value co-creation and Open Innovation can be investigated. The latter are structures that are perfectly suited to an anthropocentric society, since the participation of individuals, companies and businesses is required.

In this vein, Industry 5.0 prepares itself for important points and objectives to be met:

1. Human-Machine Collaboration: Integrating human intelligence with machine capabilities to improve creativity, problem solving and productivity.
2. Environmental Sustainability: Reduce environmental impact through the efficient use of resources, the adoption of renewable energy and the minimization of waste.
3. Mass Customization: Offering highly customized products in response to specific consumer needs, enabled by advanced technologies such as 3D printing and additive manufacturing.
4. Technological Innovation: Using emerging technologies such as Artificial Intelligence (AI), the Internet of Things (IoT), blockchain and advanced robotics to improve manufacturing processes and supply chain management.
5. Workers' well-being: Promote a safe, healthy and stimulating working environment that enhances human skills and promotes personal and professional growth.
6. Social Responsibility: Engaging in ethical and socially responsible business practices, contributing positively to the community and society.

Organizations have to achieve, at the same time, competitive advantage and corporate social responsibility within the actual socio-economical context by following these future research directions:

1. Efficiency and Process Optimization: Improve operational efficiency through the optimization of production processes and the reduction of waste.
2. Product Quality and Safety: Ensure product quality and safety through the use of advanced technologies for production monitoring and control.
3. Innovation and Competitiveness: Promote continuous innovation to maintain and improve competitiveness on the global market.
4. Sustainability and Reduction of Environmental Impact: Adopt sustainable practices to reduce environmental impact and promote the responsible use of resources.
5. Customer centrality: Putting the customer at the center of company strategies, offering products and services that meet the needs and expectations of consumers.

Trends in the Food Industry: Paying Attention to Sustainability and Future Scenarios

The food industry, being one of the most dynamic and influential sectors of the global economy, is subject to constant evolution and adaptation driven by a myriad of factors ranging from technological advancements to shifting consumer preferences. The main aim of this section is to focus on the latest trends shaping the landscape of the food industry, providing a comprehensive overview of the key forces driving change and innovation.

According to 'Statitsta' website, revenue in the food market amounts to €184.30bn in 2024 and the market is expected to grow annually by 3.64% (CAGR 2024-2028). In relation to total population figures, per person revenues of €3,071.00 are generated in 2024. In addition to that, 'Statista' also claims that in the food market, 1.5% of total revenue will be generated through online sales by 2024; that the volume is expected to amount to 43.59bn kg by 2028 and, in particular, the food market is expected to show a volume growth of 1.1% in 2025. As conclusion of this paragraph, there are two graphics that show two relevant difference between the Italian food market and the world food market.

Digitalization has emerged as a transformative force within the food industry, revolutionizing every aspect of the supply chain from production to consumption. As technology continues to advance at a rapid pace, food companies are leveraging digital solutions such as blockchain, artificial intelligence, and IoT (Internet of Things) to enhance efficiency, traceability, and consumer engagement. Blockchain technology is a decentralized digital ledger that securely and transparently records transactions and data in a way that is immutable and accessible to all network participants. In the food industry, blockchain technology is used to trace and monitor the journey of food along the entire supply chain, from production to distribution to the end, promoteing food quality and operational efficiency. It promotes greater trust and transparency between consumers and producers, contributing to the creation of a safer, more responsible, and sustainable food system. According to a recent report by McKinsey, digital technologies are projected to reshape the global food system, with 60% of food companies planning to invest in digital solutions over the next five years. These investments are expected to drive improvements in supply chain management, product quality control, and personalized consumer experiences, positioning digital-savvy companies for success in an increasingly competitive marketplace.

Consumer data and predictive analytics have become indispensable tools for food companies seeking to better understand and anticipate consumer preferences. By harnessing the power of data analytics, companies can gain valuable insights into consumer behavior, market trends, and emerging opportunities, enabling them to tailor their products and marketing strategies to meet the evolving needs of their target audience.

Research conducted by Nielsen indicates that 73% of consumers are willing to pay more for personalized food products based on their individual preferences. This highlights the growing importance of data-driven personalization in the food industry, as companies strive to deliver customized offerings that resonate with their customers on a deeper level.

Sustainability and health-consciousness have emerged as major drivers of consumer behavior, influencing purchasing decisions and shaping the strategic priorities of food companies worldwide. With an increasing emphasis on environmental stewardship and personal well-being, consumers are seeking products that are not only nutritious and delicious but also ethically sourced, environmentally friendly, and socially responsible.

According to a survey conducted by Mintel, 61% of consumers actively seek out organic food products, while 45% prefer products packaged with sustainable materials. This growing demand for sustainable and health-focused products is driving innovation in the food industry, prompting companies to explore alternative ingredients, eco-friendly packaging solutions, and transparent supply chains.

The COVID-19 pandemic has had a profound impact on the food industry, accelerating existing trends and reshaping consumer behaviors in unforeseen ways. With lockdowns, social distancing measures, and supply chain disruptions, the pandemic has forced food companies to adapt quickly to new challenges while also seizing emerging opportunities.

According to the Food and Agriculture Organization (FAO), online food sales surged by 30% in 2020, while home deliveries experienced a 45% increase as consumers shifted towards e-commerce and contactless shopping. These shifts in consumer behavior have underscored the importance of digitalization, agility, and resilience in the face of unprecedented challenges, prompting food companies to rethink their strategies and priorities in a post-pandemic world.

In the dynamic and ever-evolving landscape of the food industry, effective communication strategies stand as essential pillars for startups to not only survive but also thrive and carve out their unique identity in the market. As we embark on this exploration into the realm of communication strategies tailored specifically for food startups, we delve into a captivating journey to uncover potential future trends while shining a spotlight on the indispensable role of sustainability in shaping brand identity and positioning. The recent surge in the establishment of food startups has revolutionized the culinary landscape, presenting consumers with a diverse array of innovative products and services.

These startups, often characterized by their agility, creativity, and disruptive approach, are challenging traditional paradigms and reshaping consumer preferences. In this context, effective communication strategies emerge as indispensable tools for startups to capture the attention of their target audience, build meaningful connections, and establish a strong brand presence.

As we gaze into the horizon of the food industry, several potential future trends in communication strategies come into focus. Here I will put in evidence how hyperpersonalization, interactive experiences as well as transparency and traceability can interact with the complex sector of food start up bringing innovations.

One such trend is hyperpersonalization, wherein communication efforts are tailored to cater to the individual preferences and dietary needs of consumers. An exemplary startup embracing this trend is Freshly, an American meal kit delivery service that leverages customer data to curate personalized meal plans based on dietary preferences, allergies, and portion sizes. "Noom" stands out as a prime example of hyper-personalization in the health and wellness app sphere. Developed by Noom Inc., a company headquartered in New York City, Noom employs advanced algorithms and behavioral psychology techniques with the help of artificial intelligence to provide users with personalized support in achieving their health and fitness objectives. Through customized meal plans, individualized feedback, personalized goal setting, and activity tracking features, Noom tailors its approach to each user's preferences and needs, fostering greater engagement and facilitating more successful outcomes in their wellness journeys. Spoon Guru can be another example: Developed by Spoon Guru Ltd., a company based in London, this innovative app utilizes cutting-edge technology to cater to individuals with diverse dietary requirements, including allergies, intolerances, and specific lifestyle choices like vegetarianism or veganism. By leveraging a comprehensive database and advanced algorithms, Spoon Guru delivers personalized food recommendations, recipe suggestions, and product matches tailored to each user's unique dietary needs and preferences. This approach enhances user experience and empowers individuals to make informed food choices aligned with their health goals and dietary restrictions.

Secondly, we anticipate a rise in interactive experiences, leveraging technologies such as augmented reality (AR) and virtual reality (VR) to create immersive engagements. Picture interactive cooking tutorials in AR, guiding users step-by-step through the culinary process in their own kitchens, or virtual farm tours that provide a firsthand glimpse into the sourcing and production processes behind the food they consume. Two examples could be Kitchen Stories and Tovala: Kitchen Stories is an application developed by Kitchen Stories GmbH based in Berlin, Germany, that stands out for offering interactive experiences in the realm of cooking and recipe preparation. This culinary startup is committed to providing users not only with detailed recipes but also with vibrant culinary experience through interactive tutorials. Users can follow step-by-step instructions via engaging videos, high-quality images, and explanatory texts, making the cooking process an interactive and immersive experience. The combination of captivating visuals and detailed instructions transforms meal preparation into a playful and educational experience for Kitchen Stories users. On the other hand, Tovala a Chicago-based startup, has emerged as a pioneer in exploring interactive experiences within the culinary sphere. Founded in 2015 by David Rabie and Bryan Wilcox, Tovala offers a smart oven and meal delivery service that seamlessly integrates technology with cooking, revolutionizing the way people prepare meals at home. At the heart of Tovala's innovative approach is its smart oven, which connects to a mobile app, allowing users to control cooking functions remotely and access a library of chef-crafted recipes. This interactive platform not only simplifies.

TECHNOLOGICAL CHANGE MANAGEMENT: TRANSFORMING EXISTING MODELS AND AVOID CHAOS

The purpose of Technology Change Management is to identify, select and evaluate new technologies, tools, methods and processes and integrate them into the company organization in an orderly manner, in order to increase productivity and reduce time for the development of new products (Frankel, 1990).

In general, change is not easy, and implementing new technologies is particularly expensive; in fact, beyond the material cost, it is also necessary to take into account every internal resistance within the company, since, in general, people tend not to like changes, as they involve having to adapt to new dynamics. However, evolution is necessary to guarantee business continuity and becomes even more important when it is referred to technology, which can bring new resources to improve products, services, skills and models.

For this reason, a team dealing with technological innovation requires a special combination of personalities and skills: engineers, economists, computer scientists, with managerial skills but also logical, deductive, imagination and creativity skills.

Among the most brilliant innovations of recent years, how can we fail to mention artificial intelligence, a discipline belonging to computer science that studies how to create intelligent systems that allow the ability and behavior of individuals to be simulated (Leone et al., 2021). In this way, it would be possible to provide services that would appear to be the exclusive domain of human beings. We start from the so-called "machine learning", i.e. "automatic learning", a precious resource for the industry, as it is capable of "learning" from the data and information available; in fact, unlike computers, which perform mechanical functions and precise tasks, AI is designed to acquire and process information that allows it to learn and interact effectively with the surrounding environment.

An initial solution to propose a starting point for future research in this theme is synthesized on the following conceptual framework (Figure 2). By using a technological change management approach, the idea is to focalize the attention to:

1) Semi-open innovation
2) Sustainable business models
3) Use of collaborative robots

Figure 2. The proposed conceptual framework to sustain technological change (Author research)

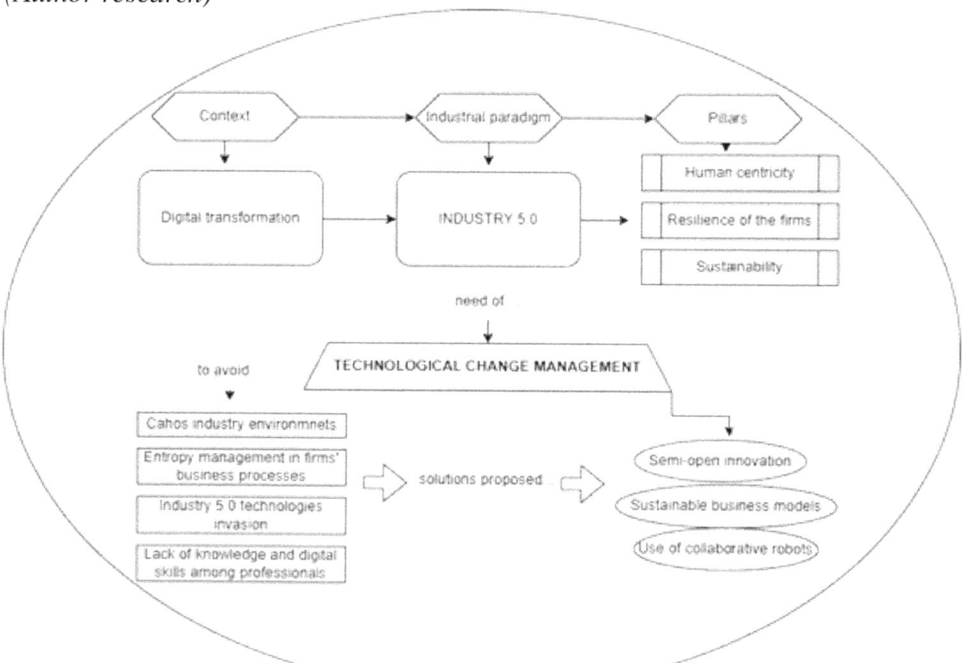

Thus, for example, Facebook exploits this type of technology to manage advertising spaces, possible "friendships" and the news to show, which are as much in line with the personal profile of the individual. Or again, the same Google search engine makes use of AI to identify, among the multitude of sites available, those most consistent with the user's request, with the influence also of the so-called "link building", which we will explore later.

Undoubtedly, data represents the "new oil" in the age of digitization, therefore AI can help companies extract them, processing billions of pieces of information in an instant, providing accurate predictions on future results, based precisely on the data historians. Clearly, Big Data would lose its usefulness if it were not possible to interpret them correctly: hence, it is still necessary that the intelligence of the machine can be added to that of people, considering the fact that the former makes decisions based only on available data, having no opinion or emotion. In light of this, in the world of marketing, data, numbers and creativity mix with technology to achieve a result that is constantly improved. Consequently, Machine learning will help marketing to better understand human behavior, analyzing the key factors of its choices: for example, with a statistical algorithm, it will be possible to identify the main causes of customer abandonment, going to act accordingly. Indeed, with the advent of new technologies, the prediction of human behavior is even more incredible: being able to understand whether or not that advertising campaign will have the desired success, whether those customers are more or less inclined to leave us and much more.

Well, the union between marketing and artificial intelligence and new technologies currently falls within areas such as segmentation and target identification, automation of communication and assistance processes, as well as web design and the creation of personalized content.

Therefore, the implementation of new technologies is a valid aid in Neuromarketing, thanks to the scientific monitoring of brain activity, eyes and other biophysical factors.

Today we talk about Emotion AI, a new branch of marketing that arises from the recognition of emotions through the use of technological devices.

Investing in Neuromarketing and artificial intelligence thus means not only better understanding the consumer, but also decreasing the chances that a new product may fail.

As part of the idea generation, it is important to understand the dynamics based on the opportunities for transformation or adaptation of existing business logic. Change has never been as rapid and invasive as it is today. Thanks to new technologies such as AI or IoT, everything dematerializes to the point that the availability and abundance of everything for everyone is a concept that should not be underestimated but managed. Traditional organizations struggle to stand up or fail, replaced by a business generation able to respond flexibly to sudden changes in consumer needs, lifestyles and tastes, but the question that arises is:

How to Transform a Business Model to Manage the Disruption That the Abundance of Resources is Causing in Various Sectors?

Transforming an existing model requires an innovative mentality towards the design of a functioning business model, that can fit into the socio-economic and cultural challenges of current markets, and that is able to respond and resist the impact of accelerating technologies. The changes in the business model, therefore, require a more in-depth analysis of global trends, as well as the ability to exploit the technologies made available by Industry 4.0. The extreme use of Cyber security, cloud computing, or the IoT is invading business processes and destroying all types of entrepreneur reasoning logic. For these reasons, we talk about technological change management in the era of digital transformation because we need to take advantage of the support of the aforementioned solutions but not be replaced or invaded in business choices.

Given these premises, let's imagine that we have to transform an existing model due to very rare events such as the pandemic we are experiencing, or because we realize that we have lost our ability to innovate, or to cope with the exhaustion of the competitive advantage or finally through the obsolescence of the business sector.

In current markets, many companies are transforming their value proposition through the design of new specific product-service solutions, with the aim of increasing revenues and building a sustainable and lasting competitive advantage over time. This transformation process is known by the term "servitization" and is strengthened with the emergence of digital technologies. Advanced services are provided to support not only the product, but also customers and their processes; or even, revolutionize their offer, replacing the sale of the product with the sale of pure service, substantially dematerializing the output of the activities. Therefore, services are transformed from an opportunity to a real source of business. All this today represents a necessity for survival and for the pursuit of sustainable competitive strategies overtime.

An example is the case of Goglio, an international group born in Italy and specializing in complete systems for packaging. Over the last few years, Goglio has transformed the business model towards flexible packaging and uses the Internet of Things to make production more efficient. This has led to the complete integration of information management; in fact, thanks to the data collected, the company is able to learn the operating conditions of the machines, and monitor actual productivity and qualitative yield. This a clear example of radical innovation towards pure service models.

The transformation cannot be motivated only with the introduction of new enabling technology, but additionally with supply innovation both in terms of product technology and processes and responsibilities of the connected service. Finally, all this translates into a profound study aimed at promoting cultural change and the business model.

As we have seen, transforming an existing model involves special challenges in creating a new logic for managing technological change. An organization will often be forced to face conflicts (e.g. relationship conflicts) and compromises between the old and the new way of doing business. Transforming the business model also means changing the organization, activities, and logics relating to the new business model may be incompatible with the status quo.

Therefore, in transforming existing models in the era of digital transformation, one must be careful to maintain a degree of negative entropy. In thermodynamics, entropy indicates a degenerative process that leads to disorder, loss of connections and system breakdown. In the managerial sphere, a degree of negative entropy must be maintained precisely in order to understand the need for what has just been saying not to happen.

For these reasons, it is natural to transpose this phenomenon to organizations that are constantly trying to improve their competitiveness and profitability by transforming their model. The goal is to avoid these invasive processes often resulting from the introduction of new technologies becoming more and more complex to manage.

The transformations of business models are inevitable since the business system is in close contact with the market system in the continuous exchange of energy: the 2 systems, therefore, tend to continuously seek new states of equilibrium. However, not all transformations have the same effects. Virtuous firms adopt structured approaches to change that allow the generation of entropy to be controlled and minimized, while others that leave this transition to chance, to the goodwill of individual decision-makers, or to the intuition of few resources, suffer from a high level of organizational disorder that generates the lengthening of the time necessary for change with a significantly higher entropy state than the previous case.

It follows that value creation and entropic reduction are two strongly interrelated concepts. The objective of the companies must not be the minimization of the degree of entropy but the control of its entity.

Obviously, the need to focus the company on digital is not in question. The point is that we need to respond to the challenges of the fourth industrial revolution through a business model appropriate to the context and that takes care of the "business potential" as the ability to be more effective in the development of certain strategic activities and thus contribute to the success or future usefulness of the organization. Some methodologies referred to by some management scholars as "Transformation and / or Exponential Organization (ExT and ExO) help to contextualize technological change to exploit opportunities and create practically new businesses or radically transform existing business lines. They are defined as exponential as they are able to grow and prosper faster than others due to the correct management of new technologies. Among others, P&G, HP, Visa, Stanley Black & Decker and Mylia (Adecco Group) have adopted and still are adopting these approaches.

Finally, the existing model can also be modified by changing the resource control asset, being careful to maintain the degree of negative entropy. For example, Uber has involved car owners who are also drivers, replacing TAXIs, so it has changed the resource control regime. The differences that the changes in control of resources can bring may be significant or insignificant but they must not have an impact on the order to be maintained within the company in order to manage future evolutions and transformations.

The LEGO case study and the long history of plastic bricks (the famous LEGO bricks), which transcend cultures and generations, can teach a lot about the importance of innovation, and experimentation but also knowing how to reinvent oneself when circumstances (or crises) require it.

For these reasons, due to these difficulties, we have seen how consolidated organizations favor an incremental form of transformation in which, once again, experimentation and learning are crucial, that is to prepare a business model and "test" if it works for the context in which you want to operate. If the model doesn't work, you need to change direction and redefine the key elements to avoid big failures. An experimental approach, therefore, makes it possible to create a more flexible system in the definition of a business and a system of knowledge that can be defined as a "guiding hypothesis", which allows us to understand the direction of the business and undertake initiatives aimed at integrating or modifying the model. The experimental approach is essential to be ready to face even the unexpected of business logic or the cases in which it is necessary to transform.

For example, for the management of a business crisis, turnaround processes could be activated towards the re-centering of core activities and core markets, perhaps with differentiated outsourcing of non-core activities in the new model to be adopted.

Alternatively, it was also found that companies are experimenting with multiple business models at the same time. Therefore, for both new and consolidated companies, the need to assess the adequacy of the business model means that it must be assessed with respect to the business context and further calibrated in order to find an optimal adaptation to the environment. It is clear that the business context has a major impact on both the creation and transformation of the business model. We can take up the thought of Adriano Olivetti, which we can connect to the transformation of the business model of a company in relation to the importance of the organization of work for the social, cultural, and human development of those who work.

"The factory was conceived on a human scale so that he could find in his tidy workplace an instrument of redemption and not a device of suffering."

EXPLORATIVE ANALYSIS AND CASE STUDIES

Treedom: A Forest of Solutions for Sustainable Businesses

What is Treedom? "There is an Eskimo in the middle of the ice, a Bedouin in the desert, and a Polynesian out at sea. All three are planting a tree. Treedom's mission is to make sure this is not a joke."

Treedom is a Benefit Corporation that, as can be inferred from its mission, seeks to combine entrepreneurial activities with environmental and social sustainability. It represents the world's first platform that allows individuals and companies to plant trees remotely and follow their growth online, financing agroforestry projects in collaboration with NGOs, local cooperatives, and farming communities in various countries around the world. Since its foundation in 2010, 1.8 million trees have been planted directly by local communities, bringing significant environmental, social, and economic benefits. Through geolocation, every user can remotely follow the life and story of their tree. This active engagement of the public makes Treedom's trees valuable marketing and communication tools for an increasing number of companies pursuing Corporate Social Responsibility (CSR) goals.

In 2014, with this business model, Treedom became part of the Certified B Corporations, a network of companies with high ethical and environmental performance.

The Story

It was in 2010 in Florence that the two friends Federico Garcea and Tommaso Speroni founded Treedom. A project born almost by chance and as a joke. They were passionate friends who used to play the video game FarmVille, where the goal was to enrich one's virtual farm with various plants and unlock new features by paying. At the same time, they were also working on a reforestation project in Cameroon. Eventually, one day, as Federico Garcea recalls, the two asked themselves: "Millions of people enjoy planting trees for fun. Why not allow them to do it for real?"

Thus, the fantastic dimension of FarmVille became a reality. Garcea and Speroni created an online platform that allows adopting a tree remotely among the various types available and following its development over the years, dispelling any doubts about using the donated money. Initially, the founders' idea was to create a plantation in Africa, but they faced initial difficulties, especially in convincing local farmers, as there was no immediate profit margin for them, since these were not fruit trees.

Over the years, the team began its climb to success with the first funding from the Tuscany region, followed by a grant of nearly half a million from the Bill & Melinda Gates Foundation. Another significant boost came from the choice of the singer Jovanotti to use the startup's services to offset the carbon dioxide emissions of his tour.

Now, Treedom is far from being the "niche project" it was initially intended to be. Ten years after its founding, around 1.8 million trees have been planted in 16 countries worldwide, including Italy. It is estimated that these trees will absorb over 360,000 tons of CO2, supporting biodiversity, sustainable resource use, active user involvement, and economic support for small rural farming communities caring for the trees.

Treedom and the Sustainable Development Goals (SDGs).

Among the ambitious goals the company sets for itself, in addition to going public, is the achievement of 10 out of the 17 Sustainable Development Goals established by the 2030 Agenda, ensuring environmental protection and prosperity for all. The impact that agroforestry projects have on communities is detailed in the link provided.

Treedom for Business Treedom also caters to companies looking for a reliable ally to effectively and engagingly communicate their commitment to sustainability. Presently, about 6,500 companies rely on the services of this platform. Among them are notable names like Samsung, Timberland, Findus, Enel, Berlucchi, and Ferragamo. These companies, from different sectors, share the common goal of creating value not only for themselves and their customers but for the entire planet.

Treedom for Business offers two packages, one specific for medium-large companies and the other for small businesses looking to step into the world of sustainability. Both provide a comprehensive campaign from conception and planning to implementation to create tailored engagement and communication strategies for each company. Treedom transforms each tree into a brand ambassador for the company and its environmental commitment, allowing companies to create a true corporate forest, share it with their employees and customers, making them feel part of a story, of a change.

As noted by Federico Fraboni, Sustainability Manager of the Calzedonia Group, the added value of collaborating with Treedom lies in its ability to have a strong emotional impact on the public. With the simple act of gifting a tree, it manages to touch and intrigue people. The benefits are made tangible by measuring the CO2 absorbed by the planted trees, sharing photos of the tree's life evolution, the farmer, and the community taking care of it. All this is done with full transparency.

" A tree absorbs CO2 and gives oxygen, it can retain soil with its roots and be a shelter for many animals, give nectar and beauty with its flowers, and finally fruits for all the farmers involved in our projects. A tree is truly the best tool we have to give a future to our planet." - Federico Garcea.

The IMA Case (Automated Solutions for Packaging Industry)

A sector particularly interested in the issue of sustainability which could be deduced from the previous chapters and which has a significant impact in the sphere of the sustainable market is the production of packaging, i.e. the packaging of the products that we use even of those products that not even we imagine the fabric wrapping that envelops our infusion due to its very small size but such a great impact as tea bags. A proud Italian world leader in the packaging sector is IMA (Automated Solutions for Packaging Industry).

It was founded in 1961 as a small business; after all, like numerous companies in the Bologna area, today it currently holds the role of world leader for the production of machines and for the packaging of cosmetic, pharmaceutical and food products. IMA was born from the idea of Andrea Romagnoli, an inventor who had a commercial propensity and tested many types of products until the solution to this research was found; at the time it was the famous tea bag. Many automatic machines were created in these IMA factories for the packaging of medicines, cosmetics, tea and coffee produced by the largest multinationals, a world leadership achieved in 50 years of activity thanks to constant commitment, development and research into new technologies. The IMA group invests at least 5% of its annual turnover in research, this means having hundreds of designers including mechanics and electronics who continuously work every year on new projects, new machines and new technologies so that the company is always at the keep pace with the market, the machines are highly customized to specific customer requests and are made differently from each other, this is because it is the product that is in demand and therefore depending on the product it is necessary to dose the speed and dosing precision in such a way as to offer a very high degree of customization. Despite the crisis, it is a company that has maintained constant growth, especially on the foreign market which represents 93% of turnover. IMA is not only mechanical or electronic but also chemists who study and think about the science of materials and manage to include different professional skills in the business plan and at the same time offer the market and this is their strong point. For this company, economic results and social benefits are complementary through which it is possible to generate value for the company; employees, suppliers and customers for the entire community but knows perfectly well that to continue towards a growth path it is important to implement a mentality with broader and above all current objectives, at this moment the priority is not only business but also the environment and the society we live in. IMA has introduced a plan towards sustainability; IMA ZERO a commitment to the creation of products, production processes and services benefiting from a logic of sustainability and the awareness that their efforts today will be able to take shape in the future.

IMA ZERO was launched in 2019 as a demonstration of IMA's commitment to sustainable development that began many years ago with the annual publication of the sustainability report. It includes all the practices already in place, plus those we are currently implementing, and has both the final objective of reducing the environmental impact of industrial production to a minimum, and that of promoting and creating a work and life habitat that always people and biodiversity are at the centre.

IMA's pioneering approach to sustainability leads to finding innovative and ecological ways to reduce energy and water consumption, offset emissions, manage waste by promoting recycling, replace plastic with more sustainable materials, promote sustainable mobility and, more still in general, to undertake more sustainable actions by orienting operations towards a circular economy.

IMA's business has allowed it to grow and compete rapidly with international realities, requiring considerable determination and consistency in evolving, implementing the corporate culture, making its values clearer and more explicit and the responsibility it intends to assume both internally and externally. Over the course of 60 years IMA has built multidisciplinary expertise, identified its values, its potential and has developed products and services based on it, investing in new technologies that have contributed to improving the quality of life by reducing food waste, creating of effective drugs and greater respect for the environment. In fact, this leader works to create wealth, well-being and defining the ethical and social responsibility of all the members of the IMA business organisation.

All these initiatives, applications and projects are possible thanks also to the constitutional charter, a means of implementing ethics within the company by managing transactions and human resources in a fair and effective way but also externally by supporting its reputation; to involve the staff in the ethical principles as much as possible, the latter are attached together with the new contracts, this also applies to suppliers, to establish commercial relationships on ethical standards. The Sustainable Development Goals identified by the United Nations IMA Group adheres to these objectives in its sustainability policy. With a view to this policy, it has mobilized in open innovation, such as marketing input, machine analysis, collaborative innovation and enabling technologies. IMA Group has activated collaborations on projects that have aroused interest in research institutes and universities in Italy but also in the spinouts of the Massachusetts Institute of Smart Machines and Factories, including the development of intelligent mechatronic systems, cognitive automation, artificial intelligence, cybersecurity for be able to apply the digitalization of the supply chain and additive manufacturing.

The company has a sustainable policy that reports the group's commitments on various sustainability issues.

The various commitments can be traced back to a few points:

- ♣ reduction of impacts
- ♣ investments in research and development for sustainable innovation
- ♣ promotion of responsible behavior along the value chain

The governance of a company is not only given by the ability or complexity of developing a "winning" strategy but also by the ability to guarantee a balance between competitiveness, sustainability and social responsibility. The IMA SPA internal control system allows you to manage, measure and monitor risks so that healthy, correct and consistent management occurs with the set objectives of sustainable business development. For its management system, IMA has equipped itself with a set of certificates aimed at creating value in a sustainable way shared with all stakeholders, first and foremost the resources.

The interpretation of the needs of its end-users is at the forefront of its objectives, pursuing the best quality standards, adopting the best practices to protect the health and safety of its workers and reducing environmental impact to a minimum. In particular, IMA has achieved and maintains consistency with a series of international certifications. Furthermore, a project to extend the corporate quality system to the division was introduced in 2022.

For the company under consideration, sustainability is not exclusively a strategy but a way of operating, it is a company promise to create value that persists over time.

IMA's sustainability policy is based on the code of ethics which defines the behaviors which are extended to all the companies that make up the IMA groups and made public. Commitments towards sustainability are divided into five macro areas:

1. Product quality and responsibility
2. Governance and business integrity
3. People
4. Environment
5. Community and knowledge development

Regarding the sustainability applied in the company, the primary objective of IMA's business consists in satisfying the end-user, i.e. the person who in the electronic sector is at the end of the production-commercial-use chain, encouraging their innovative capacity to reduce environmental impacts, increasing quality of products and safety by investing in research and development.

REFERENCES

Appio, F. P., Frattini, F., Petruzzelli, A. M., & Neirotti, P. (2021). Digital transformation and innovation management: A synthesis of existing research and an agenda for future studies. *Journal of Product Innovation Management*, *38*(1), 4–20. DOI: 10.1111/jpim.12562

Bocken, N. M., Short, S. W., Rana, P., & Evans, S. (2014). A literature and practice review to develop sustainable business model archetypes. *Journal of Cleaner Production*, *65*, 42–56. DOI: 10.1016/j.jclepro.2013.11.039

Dean, T. J., & McMullen, J. S. (2007). Toward a theory of sustainable entrepreneurship: Reducing environmental degradation through entrepreneurial action. *Journal of Business Venturing*, *22*(1), 50–76. DOI: 10.1016/j.jbusvent.2005.09.003

Evans, S., Vladimirova, D., Holgado, M., Van Fossen, K., Yang, M., Silva, E. A., & Barlow, C. Y. (2017). Business model innovation for sustainability: Towards a unified perspective for creation of sustainable business models. *Business Strategy and the Environment*, *26*(5), 597–608. DOI: 10.1002/bse.1939

Frankel, J. A. (2000). Globalization of the Economy.

Geissdoerfer, M., Vladimirova, D., & Evans, S. (2018). Sustainable business model innovation: A review. *Journal of Cleaner Production*, *198*, 401–416. DOI: 10.1016/j.jclepro.2018.06.240

Huang, M. H., & Rust, R. T. (2018). Artificial intelligence in service. *Journal of Service Research*, *21*(2), 155–172. DOI: 10.1177/1094670517752459

Kolk, A., & Pinkse, J. (2007). Towards strategic stakeholder management? Integrating perspectives on sustainability challenges such as corporate responses to climate change. *Corporate Governance (Bradford)*, *7*(4), 370–378. DOI: 10.1108/14720700710820452

Kramer, M. R., & Porter, M. (2011). *Creating shared value* (Vol. 17). FSG.

Leone, D., & Schiavone, F. (2019). Innovation and knowledge sharing in crowdfunding: How social dynamics affect project success. *Technology Analysis and Strategic Management*, *31*(7), 803–816. DOI: 10.1080/09537325.2018.1554858

Leone, D., Schiavone, F., Appio, F. P., & Chiao, B. (2021). How does artificial intelligence enable and enhance value co-creation in industrial markets? An exploratory case study in the healthcare ecosystem. *Journal of Business Research*, *129*, 849–859. DOI: 10.1016/j.jbusres.2020.11.008

Lüdeke-Freund, F. (2010). Towards a conceptual framework of'business models for sustainability'. In Wever, R., Quist, J., Tukker, A., Woudstra, J., Boons, F., & Beute, N. (Eds.), *Knowledge collaboration & learning for sustainable innovation* (pp. 25–29).

McQuade, P., & Schmitz, M. (2017). The great moderation in international capital flows: A global phenomenon? *Journal of International Money and Finance*, *73*, 188–212. DOI: 10.1016/j.jimonfin.2017.02.027

Nonaka, L., Takeuchi, H., & Umemoto, K. (1996). A theory of organizational knowledge creation. *International Journal of Technology Management*, *11*(7-8), 833–845.

Ritala, P., Huotari, P., Bocken, N., Albareda, L., & Puumalainen, K. (2018). Sustainable business model adoption among S&P 500 firms: A longitudinal content analysis study. *Journal of Cleaner Production*, *170*, 216–226. DOI: 10.1016/j.jclepro.2017.09.159

Roome, N., & Louche, C. (2016). Journeying toward business models for sustainability: A conceptual model found inside the black box of organisational transformation. *Organization & Environment*, *29*(1), 11–35. DOI: 10.1177/1086026615595084

Schaltegger, S., Lüdeke-Freund, F., & Hansen, E. G. (2016). Business models for sustainability: A co-evolutionary analysis of sustainable entrepreneurship, innovation, and transformation. *Organization & Environment*, *29*(3), 264–289. DOI: 10.1177/1086026616633272

Teece, D. J. (2000). Strategies for managing knowledge assets: the role of firm structure and industrial context. *Long range planning, 33*(1), 35-54.Itami, H., & Roehl, T. W. (1991). *Mobilizing invisible assets.* Harvard University Press.

Compilation of References

Abdelkafi, N., & Täuscher, K. (2016). Business models for sustainability from a system dynamics perspective. *Organization & Environment*, *29*(1), 74–96. DOI: 10.1177/1086026615592930

Abernathy, W. J., & Utterback, J. M. (1978). Patterns of industrial innovation. *Technology Review*, *80*(7), 40–47.

Acs, Z. J., & Audretsch, D. B. (1990). *Innovation and small firms*. MIT press. DOI: 10.1007/978-94-015-7854-7

Adams, M. (2020). The kingdom of dogs: Understanding Pavlov's experiments as human–animal relationships. *Theory & Psychology*, *30*(1), 121–141. DOI: 10.1177/0959354319895597

Adams, R., Bessant, J., & Phelps, R. (2006). Innovation management measurement: A review. *International Journal of Management Reviews*, *8*(1), 21–47. DOI: 10.1111/j.1468-2370.2006.00119.x

Akhtar, M. N., Haleem, A., & Javaid, M. (2024). Exploring the advent of Medical 4.0: A bibliometric analysis systematic review and technology adoption insights. *Informatics and Health*, *1*(1), 16–28. DOI: 10.1016/j.infoh.2023.10.001

Alexander, T. (2023). Unwrapping the McDonald's model: An introduction to dynamic social theory. *Journal of American Culture*, *46*(3).

Ali, A., Davis, E., Martinez, C., & Brown, L. (2024). The Role of Organizational Culture in Driving Innovation and Competitive Advantage. *International Journal of Management. Business Economics (Cleveland, Ohio)*, *1*(1).

Ali, M., Shujahat, M., Fatima, N., Lopes de Sousa Jabbour, A. B., Vo-Thanh, T., Salam, M. A., & Latan, H. (2024). Green HRM practices and corporate sustainability performance. *Management Decision*, *62*(11), 3681–3703. DOI: 10.1108/MD-05-2023-0787

Alon, I., Apriliyanti, I. D., & Henríquez Parodi, M. C. (2021). A systematic review of international franchising. *Multinational Business Review, 29*(1), 43–69. DOI: 10.1108/MBR-01-2020-0019

Amini, M., & Jahanbakhsh Javid, N. (2023). A multi-perspective framework established on diffusion of innovation (DOI) theory and technology, organization and environment (TOE) framework toward supply chain management system based on cloud computing technology for small and medium enterprises. *Organization and Environment (TOE) Framework Toward Supply Chain Management System Based on Cloud Computing Technology for Small and Medium Enterprises (January 2023).International Journal of Information Technology and Innovation Adoption, 11*, 1217–1234.

Appio, F. P., Frattini, F., Petruzzelli, A. M., & Neirotti, P. (2021). Digital transformation and innovation management: A synthesis of existing research and an agenda for future studies. *Journal of Product Innovation Management, 38*(1), 4–20. DOI: 10.1111/jpim.12562

Arnold, S. (2021). *Gender and Early Television: Mapping Women's Role in Emerging US and British Media, 1850-1950*. Bloomsbury Publishing. DOI: 10.5040/9781350192478

Arora, A., Fosfuri, A., & Gambardella, A. (2004). *Markets for technology: The economics of innovation and corporate strategy*. MIT press.

Audretsch, D. B., Braunerhjelm, P., & Carlsson, B. (2005). *Growth and entrepreneurship: An empirial assessment* (No. 3205). Papers on Entrepreneurship, Growth and Public Policy.

Azeem, M., Ahmed, M., Haider, S., & Sajjad, M. (2021). Expanding competitive advantage through organizational culture, knowledge sharing and organizational innovation. *Technology in Society, 66*, 101635. DOI: 10.1016/j.techsoc.2021.101635

Bastone, A., Leone, D., & Schiavone, F. (2023). Industrial Convergence and Digital Skills in Service Industries: An Explorative Analysis. *IEEE Transactions on Engineering Management, 71*, 3354–3362. DOI: 10.1109/TEM.2023.3342641

Beattie, V., & Smith, S. J. (2013). Value creation and business models: Refocusing the intellectual capital debate. *The British Accounting Review, 45*(4), 243–254. DOI: 10.1016/j.bar.2013.06.001

Becattini, G. (1999). La fioritura della piccola impresa e il ritorno dei distretti industriali. *Economia e Politica Industriale, 1999*(103).

Benyus, J. M. (1997). *Biomimicry: Innovation inspired by nature* (Vol. 688136915). Morrow.

Bocken, N. M., Schuit, C. S., & Kraaijenhagen, C. (2018). Experimenting with a circular business model: Lessons from eight cases. *Environmental Innovation and Societal Transitions, 28*, 79–95. DOI: 10.1016/j.eist.2018.02.001

Bocken, N. M., Short, S. W., Rana, P., & Evans, S. (2014). A literature and practice review to develop sustainable business model archetypes. *Journal of Cleaner Production, 65*, 42–56. DOI: 10.1016/j.jclepro.2013.11.039

Bocken, N., Boons, F., & Baldassarre, B. (2019). Sustainable business model experimentation by understanding ecologies of business models. *Journal of Cleaner Production, 208*, 1498–1512. DOI: 10.1016/j.jclepro.2018.10.159

Boer, H., Berger, A., Chapman, R., & Gertsen, F. (2017). *CI changes from suggestion box to organisational learning: Continuous improvement in Europe and Australia.* Routledge.

Bozkus, K. (2023). Organizational culture change and technology: Navigating the digital transformation. In *Organizational Culture-Cultural Change and Technology*. IntechOpen.

Bretas, V. P. G., & Alon, I. (2021). Franchising in Brazil: Foreign entrants build a presence. *The Journal of Business Strategy, 42*(6), 374–381. DOI: 10.1108/JBS-03-2020-0051

Burns, K. S. (2021). The history of social media influencers. *Research perspectives on social media influencers and brand communication*, 1-22.

Burton, I. (1987). Report on reports: Our common future: The world commission on environment and development. *Environment, 29*(5), 25–29. DOI: 10.1080/00139157.1987.9928891

Carayannis, E. G., & Campbell, D. F. (2009). 'Mode 3' and' Quadruple Helix': Toward a 21st century fractal innovation ecosystem. *International Journal of Technology Management, 46*(3-4), 201–234. DOI: 10.1504/IJTM.2009.023374

Cezarino, L. O., Liboni, L. B., Hunter, T., Pacheco, L. M., & Martins, F. P. (2022). Corporate social responsibility in emerging markets: Opportunities and challenges for sustainability integration. *Journal of Cleaner Production, 362*, 132224. DOI: 10.1016/j.jclepro.2022.132224

Chen, K. L., Brozen, M., Rollman, J. E., Ward, T., Norris, K., Gregory, K. D., & Zimmerman, F. J. (2020). Transportation access to health care during the COVID-19 pandemic: Trends and implications for significant patient populations and health care needs.

Chen, X., Liu, X., & Wang, X. (2022, November). Comparative Analysis of Disney+ and Netflix in Post-covid19 Era. In *2022 International Conference on Science Education and Art Appreciation (SEAA 2022)* (pp. 1229-1238). Atlantis Press.

Chesbrough, H. (2003). The logic of open innovation: Managing intellectual property. *California Management Review*, *45*(3), 33–58. DOI: 10.1177/000812560304500301

Christopher, M., & Towill, D. (2001). An integrated model for the design of agile supply chains. *International Journal of Physical Distribution & Logistics Management*, *31*(4), 235–246. DOI: 10.1108/09600030110394914

De Jong, A., Kabir, R., & Nguyen, T. T. (2008). Capital structure around the world: The roles of firm-and country-specific determinants. *Journal of Banking & Finance*, *32*(9), 1954–1969. DOI: 10.1016/j.jbankfin.2007.12.034

Dean, T. J., & McMullen, J. S. (2007). Toward a theory of sustainable entrepreneurship: Reducing environmental degradation through entrepreneurial action. *Journal of Business Venturing*, *22*(1), 50–76. DOI: 10.1016/j.jbusvent.2005.09.003

Dezi, L., Leone, D., Schiavone, F., & Simoni, M. (2019). The knowledge management of micro-firms in the crowd: Key challenges for successful operations. *Production Planning and Control*, *30*(10-12), 1005–1018. DOI: 10.1080/09537287.2019.1582105

Erceg, A., Bilos, A., & Kelic, I. (2021). Franchising in the digital era–What lies ahead. *Economic and Social Development: Book of Proceedings*, 53-65.

Evans, S., Vladimirova, D., Holgado, M., Van Fossen, K., Yang, M., Silva, E. A., & Barlow, C. Y. (2017). Business model innovation for sustainability: Towards a unified perspective for creation of sustainable business models. *Business Strategy and the Environment*, *26*(5), 597–608. DOI: 10.1002/bse.1939

Fatima, T., & Elbanna, S. (2023). Corporate social responsibility (CSR) implementation: A review and a research agenda towards an integrative framework. *Journal of Business Ethics*, *183*(1), 105–121. DOI: 10.1007/s10551-022-05047-8 PMID: 35125567

Felländer, A., Ingram, C., & Teigland, R. (2015). Sharing economy. In *embracing change with caution. Näringspolitiskt Forum Rapport* (Vol. 11).

Felstead, A. (2024). *The corporate paradox: Power and control in the business franchise*. Taylor & Francis. DOI: 10.4324/9781003527305

Floridi, L. (2014). *The fourth revolution: How the infosphere is reshaping human reality*. OUP Oxford.

Foxall, G. R. (1987). Radical behaviorism and consumer research theoretical promise and empirical problems. *International Journal of Research in Marketing*, *4*(2), 111–127. DOI: 10.1016/0167-8116(87)90003-6

Francisco, C. (2024). The Power of Influence: Leading by Example for Greater Role and Productivity. *Available atSSRN* 4862247.

Francisco, M. (2024). Correction: Artificial intelligence for environmental security: national, international, human and ecological perspectives (vol 61, 101250, 2023). *Current Opinion in Environmental Sustainability*, •••, 67.

Frankel, J. A. (2000). Globalization of the Economy.

Gaitán, M. G. (2021). Licensing as a Central Structure of Technology Transfer Agreements–Joint Venture and Franchising Agreements. *GRUR International*, *70*(5), 427–439. DOI: 10.1093/grurint/ikab032

Gassmann, O., Frankenberger, K., & Csik, M. (2013). Revolutionizing the business model. In *Management of the fuzzy front end of innovation* (pp. 89–97). Springer International Publishing.

Geissdoerfer, M., Vladimirova, D., & Evans, S. (2018). Sustainable business model innovation: A review. *Journal of Cleaner Production*, *198*, 401–416. DOI: 10.1016/j.jclepro.2018.06.240

Gereffi, G., Lim, H. C., & Lee, J. (2021). Trade policies, firm strategies, and adaptive reconfigurations of global value chains. *Journal of International Business Policy*, *4*(4), 506–522. DOI: 10.1057/s42214-021-00102-z

Geunes, J., & Pardalos, P. M. (2003). Network optimization in supply chain management and financial engineering: An annotated bibliography. *Networks*, *42*(2), 66–84. DOI: 10.1002/net.10082

Glaenzer, E. (2016). *Are the Brain and the Mind One?* Neuromarketing and How Consumers Make Decisions.

Golinelli, G., & Dezi, L. (1997). *Reti, finanza, progetti*. Cedam.

Gopakumar, S., & Dananjayan, M. P. (2023). Augmented reality in modern marketing: The ultimate solution to engage the digitally overwhelmed, budget-conscious audience?. *Journal of Information Technology Teaching Cases*, 20438869231202712.

Gorovaia, N., Pajić, M., & Windsperger, J. (2023). Performance of knowledge transfer mechanisms: The case of franchising networks. *Industrial Marketing Management*, *115*, 539–549. DOI: 10.1016/j.indmarman.2023.09.019

Grant, R. M. (1999). Transforming uncertainty into success: The strategic leadership forum 1999. *Strategy and Leadership*, *27*(4/5), 31–36. DOI: 10.1108/eb054645

Grech, A., Mehnen, J., & Wodehouse, A. (2023). An extended AI-experience: Industry 5.0 in creative product innovation. *Sensors (Basel)*, *23*(6), 3009. DOI: 10.3390/s23063009 PMID: 36991718

Greenhalgh, T., Robert, G., Macfarlane, F., Bate, P., & Kyriakidou, O. (2004). Diffusion of innovations in service organizations: Systematic review and recommendations. *The Milbank Quarterly*, *82*(4), 581–629. DOI: 10.1111/j.0887-378X.2004.00325.x PMID: 15595944

Hadjielias, E., Dada, O. L., & Eliades, K. (2021). Entrepreneurial process in international multiunit franchise outlets: A social capital perspective. *Journal of Business Research*, *134*, 13–28. DOI: 10.1016/j.jbusres.2021.05.022

Haeckel, S. H. (1999). *Adaptive enterprise: Creating and leading sense-and-respond organizations*. Harvard business press.

Hahn, T., Pinkse, J., Preuss, L., & Figge, F. (2015). Tensions in corporate sustainability: Towards an integrative framework. *Journal of Business Ethics*, *127*(2), 297–316. DOI: 10.1007/s10551-014-2047-5

Hammou, K. A., Galib, M. H., & Melloul, J. (2013). The contributions of neuromarketing in marketing research. *Journal of Management Research*, *5*(4), 20. DOI: 10.5296/jmr.v5i4.4023

Hannah Wren. (2024). " What is employee well-being and why is it important?", Zandesk, 16 aprile https://alleyoop.ilsole24ore.com/2024/03/13/4-days-working-week/ https://www.linkedin.com/pulse/tracing-origins-workplace-wellbeing-p6qof https://www.zendesk.com/it/blog/employee-wellbeing/

Harizi, A. (2021). Imputing Emotions to Foods.

Helmold, M. (2022). *Performance excellence in marketing, sales and pricing. Berliini*. Springer. DOI: 10.1007/978-3-031-10097-0

Henderson, R. M., & Clark, K. B. (1990). Architectural innovation: The reconfiguration of existing product technologies and the failure of established firms. *Administrative Science Quarterly*, *35*(1), 9–30. DOI: 10.2307/2393549

Hoofnagle, C. J., Van Der Sloot, B., & Borgesius, F. Z. (2019). The European Union general data protection regulation: What it is and what it means. *Information & Communications Technology Law*, *28*(1), 65–98. DOI: 10.1080/13600834.2019.1573501

Huang, M. H., & Rust, R. T. (2018). Artificial intelligence in service. *Journal of Service Research*, *21*(2), 155–172. DOI: 10.1177/1094670517752459

Iaione, C. (2016). The CO-City: Sharing, collaborating, cooperating, and commoning in the city. *American Journal of Economics and Sociology*, *75*(2), 415–455. DOI: 10.1111/ajes.12145

Itami, H., & Roehl, T. W. (1991). *Mobilizing invisible assets*. Harvard University Press.

Iyer, S. (2024). From Film to Franchise: The Rise of Cinematic Universes and Their Impact on Global Cinema. *Shodh Sagar Journal of Language, Arts. Culture and Film*, *1*(2), 14–20.

Jell-Ojobor, M., Alon, I., & Windsperger, J. (2022). The choice of master international franchising–A modified transaction cost model. *International Business Review*, *31*(2), 101942. DOI: 10.1016/j.ibusrev.2021.101942

Joshi, Y., Lim, W. M., Jagani, K., & Kumar, S. (2023). Social media influencer marketing: Foundations, trends, and ways forward. *Electronic Commerce Research*, •••, 1–55.

Kagermann, H., Lukas, W. D., & Wahlster, W. (2011). Industrie 4.0: Mit dem Internet der Dinge auf dem Weg zur 4. industriellen Revolution. *VDI nachrichten*, *13*(1), 2-3.

Kahneman, D., & Tversky, A. (2013). Prospect theory: An analysis of decision under risk. In *Handbook of the fundamentals of financial decision making: Part I* (pp. 99-127).

Karwowski, M., & Raulinajtys-Grzybek, M. (2021). The application of corporate social responsibility (CSR) actions for mitigation of environmental, social, corporate governance (ESG) and reputational risk in integrated reports. *Corporate Social Responsibility and Environmental Management*, *28*(4), 1270–1284. DOI: 10.1002/csr.2137

Kleinginna, P. R.Jr, & Kleinginna, A. M. (1988). Current trends toward convergence of the behavioristic, functional, and cognitive perspectives in experimental psychology. *The Psychological Record, 38*(3), 369–392. DOI: 10.1007/BF03395030

Klinčeková, S. (2016). Neuromarketing–research and prediction of the future. *International Journal of Management Science and Business Administration, 2*(2), 53–57.

Ko, S. (2019). *Didi Chuxing: expansion and risk management*. SAGE Publications: SAGE Business Cases Originals.

Kolar, E. (2014). *Neuromarketing and marketing management: contributions of neuroscience for the traditional marketing mix* (Bachelor's thesis, University of Twente).

Kolk, A., & Pinkse, J. (2007). Towards strategic stakeholder management? Integrating perspectives on sustainability challenges such as corporate responses to climate change. *Corporate Governance (Bradford), 7*(4), 370–378. DOI: 10.1108/14720700710820452

Kramer, M. R., & Porter, M. (2011). *Creating shared value* (Vol. 17). FSG.

La Londe, B. J., & Masters, J. M. (1994). Emerging logistics strategies: Blueprints for the next century. *International Journal of Physical Distribution & Logistics Management, 24*(7), 35–47. DOI: 10.1108/09600039410070975

Lazonick, W. (2007). The US stock market and the governance of innovative enterprise. *Industrial and Corporate Change, 16*(6), 983–1035. DOI: 10.1093/icc/dtm030

Lazzeretti, L. (2023). What is the role of culture facing the digital revolution challenge? Some reflections for a research agenda. *Rethinking Culture and Creativity in the Digital Transformation*, 10-30.

Leone, D., & Schiavone, F. (2019). Innovation and knowledge sharing in crowdfunding: How social dynamics affect project success. *Technology Analysis and Strategic Management, 31*(7), 803–816. DOI: 10.1080/09537325.2018.1554858

Leone, D., Schiavone, F., Appio, F. P., & Chiao, B. (2021). How does artificial intelligence enable and enhance value co-creation in industrial markets? An exploratory case study in the healthcare ecosystem. *Journal of Business Research, 129*, 849–859. DOI: 10.1016/j.jbusres.2020.11.008

Lescrauwaet, L., Wagner, H., Yoon, C., & Shukla, S. (2022). Adaptive legal frameworks and economic dynamics in emerging tech-nologies: Navigating the intersection for responsible innovation. *Law and Economics, 16*(3), 202–220. DOI: 10.35335/laweco.v16i3.61

Lewis, M. W. (2000). Exploring paradox: Toward a more comprehensive guide. *Academy of Management Review*, *25*(4), 760–776. DOI: 10.2307/259204

Leydesdorff, L. (2012). The triple helix, quadruple helix,..., and an N-tuple of helices: Explanatory models for analyzing the knowledge-based economy? *Journal of the Knowledge Economy*, *3*(1), 25–35. DOI: 10.1007/s13132-011-0049-4

Leydesdorff, L., & Etzkowitz, H. (1996). Emergence of a Triple Helix of university—industry—government relations. *Science & Public Policy*, *23*(5), 279–286.

Lindgreen, A., & Swaen, V. (2010). Corporate social responsibility. *International Journal of Management Reviews*, *12*(1), 1–7. DOI: 10.1111/j.1468-2370.2009.00277.x

Liu, J., & Liu, Y. (2024). Marketing charm of magic kingdom: An analysis of Disneyland's marketing strategies. In *SHS Web of Conferences* (Vol. 193, p. 01008). EDP Sciences.

Lüdeke-Freund, F. (2010). Towards a conceptual framework of'business models for sustainability'. In Wever, R., Quist, J., Tukker, A., Woudstra, J., Boons, F., & Beute, N. (Eds.), *Knowledge collaboration & learning for sustainable innovation* (pp. 25–29).

Lukito, D., Suharnomo, S., & Perdhana, M. S. (2022). Transformation management capabilities for digital transformation initiatives: A construct conceptualization in alignment with the dynamic capabilities framework. *Journal of Organizational Management Studies*, *1*, 1–22. DOI: 10.5171/2022.845443

MacArthur, E. (2013). Towards the circular economy. *Journal of Industrial Ecology*, *2*(1), 23–44.

MacVaugh, J., & Schiavone, F. (2010). Limits to the diffusion of innovation: A literature review and integrative model. *European Journal of Innovation Management*, *13*(2), 197–221. DOI: 10.1108/14601061011040258

Maines, R. (2023). Gendered Capitalism: Sewing Machines and Multinational Business in Spain and Mexico, 1850–1940 by Paula A. de la Cruz-Fernández. *Technology and Culture*, *64*(4), 1303–1305. DOI: 10.1353/tech.2023.a911013

Ma, L., Zhang, X., Ding, X., & Wang, G. (2019). Risk perception and intention to discontinue use of ride-hailing services in China: Taking the example of DiDi Chuxing. *Transportation Research Part F: Traffic Psychology and Behaviour*, *66*, 459–470. DOI: 10.1016/j.trf.2019.09.021

Mandracchia, C. M. (2021). *Transforming the Beast: The Theatre Laboratories of the "Disney Renaissance" 1984–1994* (Doctoral dissertation, University of Maryland, College Park).

Marquesan, F. F. S., Zawislak, P. A., & Camboim, G. F. (2024). The Functions of the Proto-entrepreneur in Building up Startup Innovation Capabilities. *BAR - Brazilian Administration Review*, *21*(4), e230076. DOI: 10.1590/1807-7692bar2024230076

Mascarenhas, O. A., Kesavan, R., & Bernacchi, M. (2006). Lasting customer loyalty: A total customer experience approach. *Journal of Consumer Marketing*, *23*(7), 397–405. DOI: 10.1108/07363760610712939

McQuade, P., & Schmitz, M. (2017). The great moderation in international capital flows: A global phenomenon? *Journal of International Money and Finance*, *73*, 188–212. DOI: 10.1016/j.jimonfin.2017.02.027

Medoff, N. J., & Kaye, B. K. (2021). *Now media: The evolution of electronic communication*. Routledge. DOI: 10.4324/9781003020721

Morin, C. (2011). Neuromarketing: The new science of consumer behavior. *Society*, *48*(2), 131–135. DOI: 10.1007/s12115-010-9408-1

Moulton, C. (2022). *Having an Experience: Media Franchises, Events, and Participatory Culture* (Doctoral dissertation, Northwestern University).

Nahavandi, S. (2019). Industry 5.0—A human-centric solution. *Sustainability (Basel)*, *11*(16), 4371. DOI: 10.3390/su11164371

Nambisan, S., Lyytinen, K., Majchrzak, A., & Song, M. (2017). Digital innovation management. *Management Information Systems Quarterly*, *41*(1), 223–238. DOI: 10.25300/MISQ/2017/41:1.03

Naveed, M., Akhtar, S., & Siraj, D. (2025). Navigating the Digital Shift: Analyzing Teachers' Perceptions of Transitioning to Online Platforms at Higher Education. *iRASD. The Journal of Educational Research*, *6*(1), 10–21.

Naveed, R. T., Alhaidan, H., Al Halbusi, H., & Al-Swidi, A. K. (2022). Do organizations really evolve? The critical link between organizational culture and organizational innovation toward organizational effectiveness: Pivotal role of organizational resistance. *Journal of Innovation & Knowledge*, *7*(2), 100178. DOI: 10.1016/j.jik.2022.100178

Nonaka, L., Takeuchi, H., & Umemoto, K. (1996). A theory of organizational knowledge creation. *International Journal of Technology Management*, *11*(7-8), 833–845.

Nosratabadi, S., Mosavi, A., Shamshirband, S., Zavadskas, E. K., Rakotonirainy, A., & Chau, K. W. (2019). Sustainable business models: A review. *Sustainability (Basel), 11*(6), 1663. DOI: 10.3390/su11061663

Nyambane, R. (2021). The future of the printed book in the era of technological advancement: An imperative for digital innovation and engagement. *Journal of Information. Communication and Ethics in Society, 19*(4), 537–559. DOI: 10.1108/JICES-10-2020-0106

Nyoni, T., & Bonga, W. G. (2017). Neuromarketing: No brain, no gain!. *Dynamic Research Journals' Journal of Economics and Finance (DRJ-JEF), 2*(2), 17-29.

Omachonu, V. K., & Einspruch, N. G. (2010). Innovation in healthcare delivery systems: A conceptual framework. *The Innovation Journal, 15*(1), 1–20.

Osterwalder, A. (2004). *The business model ontology a proposition in a design science approach* (Doctoral dissertation, Université de Lausanne, Faculté des hautes études commerciales).

Osterwalder, A., Pigneur, Y., & Tucci, C. L. (2005). Clarifying business models: Origins, present, and future of the concept. *Communications of the Association for Information Systems, 16*(1), 1. DOI: 10.17705/1CAIS.01601

Ottolia, A. (2017). *Big Data e innovazione computazionale* (Vol. 28). G Giappichelli Editore.

Penrose, E. (1959). *The theory of the growth of the firm*. Blackwell.

Penrose, E. T. (2009). *The Theory of the Growth of the Firm*. Oxford university press.

Peter, F. (1954). Drucker. *The Practice of Management*, 72.

Priadi, R., & Thariq, M. (2023). Reconstruction of Communicative Rationality: A Study on the Digital Mass Media Society in Indonesia. *Journal of Law and Sustainable Development, 11*(4), e579–e579. DOI: 10.55908/sdgs.v11i4.579

Rachmad, Y. E. (2022). Revolutionizing Global Business: The Impact of Management 5.0 on International Corporate Strategies.

Raha, A., & Hajdini, I. (2023). Franchisees with multiple stakeholder roles: perceptions and conflict in franchise networks. In *Relationship Marketing in Franchising and Retailing* (pp. 48–67). Routledge. DOI: 10.4324/9781003364351-4

Rashid, A., Asif, F. M., Krajnik, P., & Nicolescu, C. M. (2013). Resource conservative manufacturing: An essential change in business and technology paradigm for sustainable manufacturing. *Journal of Cleaner Production*, *57*, 166–177. DOI: 10.1016/j.jclepro.2013.06.012

Rawnaque, F. S., Rahman, K. M., Anwar, S. F., Vaidyanathan, R., Chau, T., Sarker, F., & Mamun, K. A. A. (2020). Technological advancements and opportunities in Neuromarketing: A systematic review. *Brain Informatics*, *7*(1), 1–19. DOI: 10.1186/s40708-020-00109-x PMID: 32955675

Rice, J., Raziq, M. M., Martin, N., Fieger, P., & Rice, B. (2023). The debt crisis and the adoption of Asset-Light and Fee-Orientated (ALFO) arrangements at Marriott: 1980-1995. *Journal of Tourism. Heritage & Services Marketing*, *9*(1), 58–66.

Ritala, P., Huotari, P., Bocken, N., Albareda, L., & Puumalainen, K. (2018). Sustainable business model adoption among S&P 500 firms: A longitudinal content analysis study. *Journal of Cleaner Production*, *170*, 216–226. DOI: 10.1016/j.jclepro.2017.09.159

Rogers, E. (1962). *Diffusion of Innovations*. Free Press.

Rogers, E. M. (2010). *Diffusion of innovations*. Simon and Schuster.

Roome, N., & Louche, C. (2016). Journeying toward business models for sustainability: A conceptual model found inside the black box of organisational transformation. *Organization & Environment*, *29*(1), 11–35. DOI: 10.1177/1086026615595084

Rullani, E. (2004). La fabbrica dell'immateriale. Produrre valore con la conoscenza.

Rullani, E. (1999). Crescita e successione: La metamorfosi del capitalismo personale. *CUOA Rivista*, *1*, 13–20.

Rullani, E. (1999). La diversità italiana alla prova del postfordismo. *Economia e Politica Industriale*, *1999*(103).

Saric, D., & Mikolasik, M. (2021). The impact of electronic services on traditional services. *Developments in Information & Knowledge Management for Business Applications*, *2*, 305–334. DOI: 10.1007/978-3-030-76632-0_11

Scanagatta, C. (2021). Exploring the consumer's brain: a neuromarketing approach.

Schaltegger, S., Lüdeke-Freund, F., & Hansen, E. G. (2016). Business models for sustainability: A co-evolutionary analysis of sustainable entrepreneurship, innovation, and transformation. *Organization & Environment*, *29*(3), 264–289. DOI: 10.1177/1086026616633272

Schiavone, F., Leone, D., Caporuscio, A., & Kumar, A. (2022). Revealing the role of intellectual capital in digitalized health networks. A meso-level analysis for building and monitoring a KPI dashboard. *Technological Forecasting and Social Change*, *175*, 121325. DOI: 10.1016/j.techfore.2021.121325

Schneider, H. (2017). *Creative destruction and the sharing economy: Uber as disruptive innovation*. Edward Elgar Publishing. DOI: 10.4337/9781786433435

Schumpeter, J. A. (1934). *The Theory of Economic Development*. Harvard University Press Cambridge.

Selznick, P. (1957). *Leadership in administration: A sociological interpretation*.

Shao, H., Jin, C., Xu, J., Zhong, Y., & Xu, B. (2022). Supply-demand matching of medical services at a city level under the background of hierarchical diagnosis and treatment-based on Didi Chuxing Data in Haikou, China. *BMC Health Services Research*, *22*(1), 354. DOI: 10.1186/s12913-022-07762-4 PMID: 35300679

Sheoran, S., & Vij, S. (2023). A consumer-centric paradigm shift in business environment with the evolution of the Internet of Things: A literature review. *Vision (Basel)*, *27*(4), 431–442. DOI: 10.1177/09722629211033944

Shilling, C. (2004). The body in culture, technology and society.

Slater, S. F., & Narver, J. C. (1994). Does competitive environment moderate the market orientation-performance relationship? *Journal of Marketing*, *58*(1), 46–55. DOI: 10.1177/002224299405800104

Spence, C. (2019). Neuroscience-inspired design: From academic neuromarketing to commercially relevant research. *Organizational Research Methods*, *22*(1), 275–298. DOI: 10.1177/1094428116672003

Srinivasa, R. (2022). *Effectiveness of neuromarketing in food companies: hijacking brains* (Doctoral dissertation, Dublin, National College of Ireland).

Stalmachova, K., Chinoracky, R., & Strenitzerova, M. (2021). Changes in business models caused by digital transformation and the COVID-19 pandemic and possibilities of their measurement—Case study. *Sustainability (Basel)*, *14*(1), 127. DOI: 10.3390/su14010127

Stephany, A. (2015). *The business of sharing: Making it in the new sharing economy*. Springer. DOI: 10.1057/9781137376183

Taha, T. M. A. M. (2018). Competitive Analysis of the Global Oil and Gas Industry using Porters Five Forces Model.

Tang, Q., Xie, E., & Reddy, K. S. (2022). Global production aspirations and internationalization by state-owned enterprises: A Co-evolutionary view of state-driven industrialized economy. *International Journal of Innovation Studies*, *6*(4), 276–285. DOI: 10.1016/j.ijis.2022.08.003

Teece, D. J. (2000). Strategies for managing knowledge assets: the role of firm structure and industrial context. *Long range planning, 33*(1), 35-54.Itami, H., & Roehl, T. W. (1991). *Mobilizing invisible assets*. Harvard University Press.

Teece, D. J. (2000). Strategies for managing knowledge assets: The role of firm structure and industrial context. *Long Range Planning*, *33*(1), 35–54. DOI: 10.1016/S0024-6301(99)00117-X

Thompson, R. F. (1967). Foundations of physiological psychology.

Tian, J. (2024). The Rise and Bypassing of Streaming Media. *Advances in Economics. Management and Political Sciences*, *57*, 189–195.

Tucker, D. M., Derryberry, D., Luu, P., & Phan, K. L. (2000). Anatomy and physiology of human emotion: Vertical integration of brainstem, limbic, and cortical systems. *The neuropsychology of emotion, 14*, 56-79.

Tucker, E., & Bartkiw, T. J. (2022). Franchising and the extraction of surplus value: Excavating the legal boundary between franchisees and employees. *JL & Pol. Econ.*, *3*, 484.

Ulman, Y. I., Cakar, T., & Yildiz, G. (2015). Ethical issues in neuromarketing:"I consume, therefore I am!". *Science and Engineering Ethics*, *21*(5), 1271–1284. DOI: 10.1007/s11948-014-9581-5 PMID: 25150848

Unnava, V., & Aravindakshan, A. (2021). How does consumer engagement evolve when brands post across multiple social media? *Journal of the Academy of Marketing Science*, *49*(5), 864–881. DOI: 10.1007/s11747-021-00785-z

Upton, M. (1996). *The economics of tropical farming systems*. Cambridge University Press. DOI: 10.1017/CBO9781139172868

Utriainen, T. (2020). Neuromarketing and consumer neuroscience–the evolution and current state of the Art, an integrative review.

Van de Ven, A. H. (1986). Central problems in the management of innovation. *Management Science*, *32*(5), 590–607. DOI: 10.1287/mnsc.32.5.590

Vargo, S. L., & Lusch, R. F. (2004). Evolving to a new dominant logic for marketing. *Journal of Marketing*, *68*(1), 1–17. DOI: 10.1509/jmkg.68.1.1.24036

Varkey, P., Horne, A., & Bennet, K. E. (2008). Innovation in health care: A primer. *American Journal of Medical Quality, 23*(5), 382–388. DOI: 10.1177/1062860608317695 PMID: 18820143

Vassolo, R. S., Weisz, N., & Laker, B. (2024). The Digital revolution: harnessing the power of new technologies. In *Advanced Strategic Management: A Dynamic Approach to Competition* (pp. 257–273). Springer Nature Switzerland. DOI: 10.1007/978-3-031-57419-1_20

Vecchiato, G., Cherubino, P., Trettel, A., & Babiloni, F. (2013). *Neuroelectrical brain imaging tools for the study of the efficacy of TV advertising stimuli and their application to neuromarketing* (Vol. 3). Springer. DOI: 10.1007/978-3-642-38064-8

Webster, E. (2004). Firms' decisions to innovate and innovation routines. *Economics of Innovation and New Technology, 13*(8), 733–745. DOI: 10.1080/1043859041000 1686824

Williams, L. D. (2021). Concepts of Digital Economy and Industry 4.0 in Intelligent and information systems. *International Journal of Intelligent Networks, 2*, 122–129. DOI: 10.1016/j.ijin.2021.09.002

Yfantidou, I. (2018). A neuromarketing perspective of green advertising. The influence of environmental advertising appeals and CSR to consumers. *Aristotle University of Thessaloniki, available at:*https://ikee. lib. auth. gr/record/297480/files/GRI-2018-21381. pdf*(accessed 26 February 2019).*

Zhu, W. (2024). Decoding Disney's Marketing Mastery: A Strategic Analysis. In *SHS Web of Conferences* (Vol. 188, p. 03011). EDP Sciences. DOI: 10.1051/shsconf/202418803011

Zott, C., & Amit, R. (2010). Business model design: An activity system perspective. *Long Range Planning, 43*(2-3), 216–226. DOI: 10.1016/j.lrp.2009.07.004

About the Author

Daniele Leone, PhD, is Associate Professor at the Department of Management and Quantitative studies, University of Naples Parthenope, Italy. He received his doctoral degree in management from the University of Naples Federico II in February 2019. He was Visiting Scholar at the Norwich Business School, University of East Anglia, Norwich, UK and at the EMLV Business School, Paris, FR. He is also a member of the editorial board of the International Jour-nal of Globalisation and Small Business. His research has been published in Journal of Business Research, Technological Forecasting and Social Change, IEEE Transactions on Engineering Management, International Entrepreneur-ship and Management Journal; Journal of Intellectual Capital, Business Pro-cess Management Journal, Production, Planning & Control, Technology Anal-ysis and Strategic Management. His main research areas are technological change management, sustainable business models, crowdfunding and healthcare management. He is also working as Guest Editor for Special Issues for Journal of Business & Industrial Marketing and European Journal of Inno-vation Management.

Index

B

C

D

E

I

K

M

N

P

R

www.ingramcontent.com/pod-product-compliance
Lightning Source LLC
LaVergne TN
LVHW081326300926
849161LV00008B/1890